Lecture Notes in Computer Science 5006

Commenced Publication in 1973
Founding and Former Series Editors:
Gerhard Goos, Juris Hartmanis, and Jan van Leeuwen

T0223298

Ryszard Kowalczyk   Michael Huhns
Matthias Klusch   Zakaria Maamar
Quoc Bao Vo (Eds.)

Service-Oriented Computing: Agents, Semantics, and Engineering

AAMAS 2008 International Workshop, SOCASE 2008
Estoril, Portugal, May 12, 2008
Proceedings

 Springer

Volume Editors

Ryszard Kowalczyk
Swinburne University of Technology
Hawthorn, VIC 3122, Australia
E-mail: rkowalczyk@ict.swin.edu.au

Michael Huhns
University of South Carolina
Columbia, SC 29208, USA
E-mail: huhns@sc.edu

Matthias Klusch
German Research Center
for Artificial Intelligence (DFKI GmbH)
66123 Saarbruecken, Germany
E-mail: klusch@dfki.de

Zakaria Maamar
Zayed University
Dubai, United Arab Emirates
E-mail: zakaria.maamar@zu.ac.ae

Quoc Bao Vo
Swinburne University of Technology
Hawthorn, VIC 3122, Australia
E-mail: bvo@ict.swin.edu.au

Library of Congress Control Number: 2008926218

CR Subject Classification (1998): H.3.5, H.3.3, H.3-4, I.2, C.2.4

LNCS Sublibrary: SL 3 – Information Systems and Application, incl. Internet/Web and HCI

ISSN   0302-9743
ISBN-10   3-540-79967-2 Springer Berlin Heidelberg New York
ISBN-13   978-3-540-79967-2 Springer Berlin Heidelberg New York

Typesetting: Camera-ready by author, data conversion by Scientific Publishing Services, Chennai, India
Printed on acid-free paper   SPIN: 12269228   06/3180   5 4 3 2 1 0

Preface

The global trend towards more flexible and dynamic business process integration and automation has led to a convergence of interests between service-oriented computing, semantic technology, and intelligent multiagent systems. In particular the areas of service-oriented computing and semantic technology offer much interest to the multiagent system community, including similarities in system architectures and provision processes, powerful tools, and the focus on issues such as quality of service, security, and reliability. Similarly, techniques developed in the multiagent systems and semantic technology promise to have a strong impact on the fast-growing service-oriented computing technology.

Service-oriented computing has emerged as an established paradigm for distributed computing and e-business processing. It utilizes services as fundamental building blocks to enable the development of agile networks of collaborating business applications distributed within and across organizational boundaries. Services are self-contained, platform-independent software components that can be described, published, discovered, orchestrated, and deployed for the purpose of developing distributed applications across large heterogeneous networks such as the Internet.

Multiagent systems are also aimed at the development of distributed applications, however, from a different but complementary perspective. Service-oriented paradigms are mainly focused on syntactical and declarative definitions of software components, their interfaces, communication channels, and capabilities with the aim of creating interoperable and reliable infrastructures. In contrast, multiagent systems center on the development of reasoning and planning capabilities of autonomous problem solvers that apply behavioral concepts such as interaction, collaboration, or negotiation in order to create flexible and fault-tolerant distributed systems for dynamic and uncertain environments.

Semantic technology offers a semantic foundation for interactions among agents and services, forming the basis upon which machine-understandable service descriptions can be obtained, and, as a result, autonomic coordination among agents is made possible. On the other hand, ontology-related technologies, ontology matching, learning, and automatic generation, etc., not only gain in potential power when used by agents, but also are meaningful only when adopted in real applications in areas such as service-oriented computing.

This volume consists of the proceedings of the Service-Oriented Computing: Agents, Semantics, and Engineering (SOCASE 2008) workshop held at the International Joint Conferences on Autonomous Agents and Multiagent Systems (AAMAS 2008). The papers in this volume cover a range of topics at the intersection of service-oriented computing, semantic technology, and intelligent multiagent systems, such as: service description and discovery; planning, composition and negotiation; semantic processes and service agents; and applications.

The workshop organizers would like to thank all members of the Program Committee for their excellent work, effort, and support in ensuring the high-quality program and successful outcome of the SOCASE 2008 workshop. We would also like to thank Springer for their cooperation and help in putting this volume together.

May 2008

Ryszard Kowalczyk
Michael Huhns
Matthias Klusch
Zakaria Maamar
Quoc Bao Vo

Organization

SOCASE 2008 was held in conjunction with the 7th International Joint Conference on Autonomous Agents and Multiagent Systems (AAMAS 2008) on May 12, 2008 in Estoril, Portugal.

Organizing Committee

Ryszard Kowalczyk, Swinburne University of Technology, Australia
Michael Huhns, University of South Carolina, USA
Matthias Klusch, German Research Center for Artificial Intelligence, Germany
Zakaria Maamar, Zayed University Dubai, United Arab Emirates
Quoc Bao Vo, Swinburne University of Technology, Australia

Program Committee

Stanislaw Ambroszkiewicz, Polish Academy of Sciences, Poland
Youcef Baghdadi, Sultan Qaboos University, Oman
Djamal Benslimane, Lyon 1 University, France
Jamal Bentahar, Concordia University, Canada
Brian M. Blake, Georgetown University, USA
Peter Braun, The Agent Factory GmbH, Germany
Paul A. Buhler, College of Charleston, USA
Bernard Burg, Panasonic Research, USA
Jiangbo Dang, Siemens Corporate Research, USA
Ian Dickinson, HP Laboratories Bristol, UK
Manuel Nunez Garcia, Universidad Complutense de Madrid, Spain
Mauro Gaspari, University of Bologna, Italy
Karthik Gomadam, University of Georgia, USA
Dominic Greenwood, Whitestein Technologies, Switzerland
Jingshan Huang, University of South Carolina, USA
Margaret Lyell, Intelligent Automation, USA
Michael Mrissa, Namur University, Belgium
Ingo Mueller, Swinburne University, Australia
N.C. Narendra, IBM India Research Lab, India
Xuan Thang Nguyen, Swinburne University, Australia
Leo Obrst, The MITRE Corporation, USA
Julian A. Padget, University of Bath, UK
Maurice Pagnucco, University of New South Wales, Australia
Pavel Shvaiko, University of Trento, Italy
Giovanna Petrone, University of Turin, Italy
Debbie Richards, Macquarie University, Australia

Marwan Sabbouh, The MITRE Corporation, USA
Francisco Garca Snchez, University of Murcia, Spain
Quan Z. Sheng, University of Adelaide, Australia
Hiroki Suguri, Communication Technologies (Comtec), Japan
Jie Tang, Tsinghua University, China
Rainer Unland, University of Duisburg-Essen, Germany
Steve Wilmott, Universitat Politecnica de Catalunya, Spain
Hamdi Yahyaoui, Sharjah University, UAE

Table of Contents

A Middleware Architecture for Building Contract-Aware Agent-Based Services

Roberto Confalonieri, Sergio Álvarez-Napagao, Sofia Panagiotidi,
Javier Vázquez-Salceda, and Steven Willmott

Universitat Politècnica de Catalunya
Dept. Llenguatges i Sistemes Informàtics
C/ Jordi Girona Salgado 1-3
E - 08034 Barcelona
{confalonieri,salvarez,panagiotidi,jvazquez,steve}@lsi.upc.edu
http://www.lsi.upc.edu/~webia/KEMLG

Abstract. This paper presents a middleware to help designers in the implementation of contract-aware agent-based services. The middleware provides several components, including a contract manager, a communication manager and a workflow manager, which combine to allow agents to manage contracts and the actions associated with them. The middleware is built as part of a Web service implementation of the IST-CONTRACT framework. An electronic commerce example is used to illustrate how the components of the middleware facilitates the management and execution of agreements in a contract at run-time.

1 Introduction

One of today's major trends in electronic business technologies is the increasing adoption of Service Oriented Architectures (SOA) in general and Web services in particular as a means to provide increased automation, interoperability and flexibility in deployed systems. Recently work has seen the emergence of (contractual) *agreements* as part of the specification of the expected behaviour in a distributed business process. In particular, recent initiatives have focused on the specification of choreography patterns (such as ebXML [1]) and sets of service-level agreements (SLAs) [2], which define a set of computer-observable parameters, also called metrics [3] to be evaluated. While both of these approaches show promise, the former approach lacks flexibility - relying on rigid pre-defined patterns and the latter limits the expressiveness of (contractual) agreements to limited conditions over the values of a small set of parameters.

In the approach take in the IST-CONTRACT project, we propose a move to a more flexible contracting mechanism for Service Oriented Systems based on the following three main elements:

- The introduction of *intentional semantics* within the communication between services (based on the use of performatives such as request, inform, or commit). This is important as it re-enforces the link of actual and intended behaviour.

R. Kowalczyk et al. (Eds.): SOCASE 2008, LNCS 5006, pp. 1–14, 2008.

- The creation of a *contracting language* able not only to express a set of intended behaviours on which parties agree but also to define the way that contracts and contract-related events are negotiated and communicated.
- The creation of *higher-level behavioural control mechanisms*, centered not on the tracking of a limited set of metric values but on the monitoring of higher-level objects such as commitments, obligations and violations which can be extracted from the communication semantics.

The combination of these three elements makes it possible to monitor the behaviour of a set of actors by keeping track of the fulfillment of the agreements between them.

This paper provides a description of a set of middleware components required for the creation of such contract-aware agent-oriented services. Section 2 provides a short description of the contract framework, and the contracting language. Section 3 then describes the agent-based contract environment on which the middleware is based, and our proposal of the middleware architecture supporting contract-aware agent-based services. Section 4 shows how the middleware components work during execution. Section 5 covers related work. Finally Section 6 presents conclusions and provides directions for future work. Throughout the paper we will use a contract between a buyer and an on-line music store to exemplify how the approach works. The middleware in question is implemented in early prototype form and is currently being extended in the context of the European Commission 6th Framework project CONTRACT.[1]

2   IST-Contract Framework

The aim of the IST-CONTRACT project is to develop a framework and its infrastructure which make possible to model, build, verify and monitor distributed electronic business systems on the basis of dynamically-generated, cross-organisational contracts. Such contracts are explicit descriptions of the expected behaviours of individual services and the system as a whole.

The following sections summarise the elements the middleware is based on: the *contract theoretical framework*, and the *contracting language*.

2.1   Theoretical Framework

The IST-CONTRACT theoretical framework [4] defines a framework for the conceptualisation of contracts, of the agents dealing with these contracts and an infrastructure for supporting the management of contract-related processes.

The main concepts of the framework are *contracts* and *agents*. A *contract* document formally captures a mutual agreement between two or more agents. A contract describes a set of intended behaviours by means of *deontic clauses* (obligations, permissions, prohibitions) and the agents responsible for those clauses. The agents assigned to contract clauses are the *parties* of that contract, the *contract parties*. Contract parties are classified as either *business contract parties* or

---

[1] http://www.ist-contract.org

administrative contract parties [4]. Business contract parties are essentially the parties involved in contract-related, application specific interactions. Administrative contract parties are agent facilitation roles such as *observer, manager, notary*, and *contract storer*. In short, *observer, manager*, and *notary* roles support the monitoring of contract execution and certification of contract creation and evolution (notaries are used when the contract parties are not fully trusted). The *contract storer* keeps knowledge about contracts and provides interfaces for accessing a contract repository.

The framework is defined in a way that it can be instantiated in various technology settings. Section 3.1 briefly describes the contracting environment that has been implemented for agent-oriented Web services.

2.2   Contracting Language

The contracting language defined in [6] describes the way contracts are structured, communicated and managed. Not only does this define the format for expressing contracts but it also provides the way to make statements about such contracts, the way to structure messages and the use of protocols.

The framework is structured into 6 layers of communication (Figure 1). The *Domain Ontology Layer* contains the domain ontology, providing ontological definitions of terms, predicates and actions. The *Contract Layer* defines the contract document structure[2], which includes deontic statements about the parties' obligations, permissions and prohibitions in terms of predicates and actions defined in the previous layer. The *Message Content Layer* defines message content, allowing agents to express statements about contracts (e.g. a contract being active/inactive), and actions related to contracts (e.g. accept, sign, cancel a contract). The *Message Layer* allows agents to express their attitudes towards the content of the message by means of performatives (e.g. an agent proposes to sign contract C1, or an agent requests cancellation of a contract C2). The *Interaction Protocol Layer*, pre-defines contract handling protocols as acceptable sequences of messages to achieve a given goal (e.g. a protocol for agreeing on contract termination). Finally the *Context Layer* describes the interaction context where contracting parties will carry out the obligations, permissions and prohibitions.

Due to space restrictions, in this paper we will express the deontic expressions in the example using a human-readable notation. The full description of the contracting language including the XML structure for contract documents can be found in [6].

3   A Middleware for Contract-Aware Agents

The contract framework and contracting language described in the previous section outline the conceptualisation of contracts, the agents operating with these contracts, how contracts are specified and the way agents communicate at

---

[2] The full XML schema defining the structure of our contracts is accessible at http://www.ist-contract.org/schemas/ISTContract.xsd

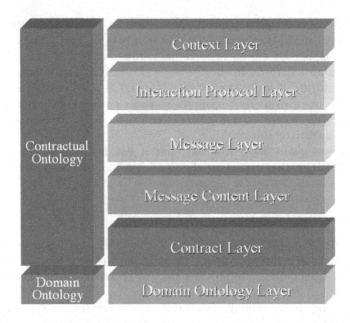

Fig. 1. The contracting language layers [17]

knowledge level about them. In practice, the framework and the agents can be instantiated in a *contracting deployment* when delivering real business applications. This leads to two main issues: the creation of the *contracting environment* the agents operate in, and the creation of the *agent middleware service architecture*, a set of components easing the creation of contract-aware agent-oriented services. In the following sections we cover both aspects, focusing especially on the internal components of the middleware.

3.1   Contracting Environment

The contracting environment is an instantiation of the theoretical framework [4] which defines business contract parties as agent-oriented Web services [7], involved in contract-related processes. Each agent may enact one or more roles (e.g. a *business contract party* or both an *observer* and *manager* for example), including administratives roles as agreed by the parties to a contract. Figure 3 shows an example of two business contract parties, a *musicBuyer* and a *music-Seller*. Administrative contract parties' roles are all enacted by extra agents: a *Notary* (enacting the notary role), and a *Conflict Manager*, enacting both the observer and the manager roles in the framework. The figure also shows how the contracting environment provides additional support systems such as a *Contract Repository*, an *Ontology Store* which ease the development of distributed contract-based applications. Additional supporting systems such as directory service (acting as agents yellow page) and context service (providing interaction context rules) can be added to the environment as well.

In the following we present our proposal of the intended agent service architecture for building contract-aware agent-based services.

3.2   Agent Middleware Service Architecture

Our proposal for the internal architecture of the agent service middleware is shown in Figure 2. Different public interfaces allow contract-aware agents to communicate with each other and with other components of the contracting environment. Public interfaces are of three types:

- *agent-to-agent:* is a Web service interface which allows agents to communicate at the knowledge level according to the contracting language and protocols defined in [6]. FIPA-ACL performatives are exchanged as Web service messages using SOAP.
- *agent-to-system:* is a Web service interface used for communication between the agent and supporting systems such as the Ontology Store and the Contract Repository.
- *agent-to-user:* is a Web interface through which the contract-aware agent interacts with system users (human interface) and with the real world. The former facilitates interactivity with the user, the latter interoperability with third party systems (such as Merchant provider for the on-line payment).

Architecturally, a contract-aware agent is split into several modules which provide specific functionalities and knowledge. The modules communicate through Java APIs.

The **decision maker** contains the core intelligence and main reasoning cycle of the agent. Although the implementation of the decision maker is left to the designer, a decision maker template is provided, which implements a practical reasoning engine.[3] This would allow to program the agent with contract-related behaviours, which model the deliberation and the achievement about contract clauses. To implement contract-related behaviours, the decision maker is assisted by the *contract manager*, the *communication manager* and *workflow manager*. These modules not only handle the low level details of messages, contract agreements and workflows but also keep an up-to-date model of the status of all these aspects which can be easily queried by the decision maker. This separation of concerns eases the implementation from the designer perspective, as the reasoning cycle inside the decision maker focuses on strategic decision-making.

The **contract manager** contains the contract knowledge and the business logic of the contracts in terms of predicates and actions (which are formally defined in the *Ontology Store*). The contract manager is aware of the contract deontic clauses that apply to the given agent, their status (active, inactive, violated), and the overall status of the contract. The contract manager also keeps track of pending obligations and signals the decision maker module about next obligations to be achieved, risks of violating a clause or the fulfillment of the contract.

---

[3] The reasoning is directed toward actions based on beliefs, desires, and intentions [5].

The **workflow manager** contains the operational knowledge required to execute and monitor the contract workflow, i.e. the order in which the actions associated to the deontic clauses of the contract have to be carried on. Operational knowledge is expressed in terms of action's inputs, outputs, and pre- and postconditions. Pre- and post-conditions straightforwardly map to activating and terminating conditions of contract deontic clauses. In this way, a workflow execution results in the execution of a sequence of actions which eventually satisfy contract obligations. Fulfilled sequences are signalled to the contract manager, which marks active obligations as fulfilled and forwards the status change to the decision maker module, which can deliberate then on the next step to do.

The **communication manager** handles all issues related to agent-to-agent communication. One of its functionalities is to act as a local Directory Facilitator for the business agent, being aware of the network topology, i.e. it knows which are the active agents in the system and how to contact them. Secondly it knows which interaction protocols to choose for contract-related behaviours. E.g., during the process of setting a new contract, the communication manager is queried by the decision maker about the protocol to use to achieve a communication goal (setting the contract). The communication manager is assisted by two modules: the *dialogue* and *message managers*.

The **dialogue manager** implements a library of interaction protocols, storing the sets of acceptable sequences of messages which fulfill the goal state of a protocol for each of them. The dialogue manager is implemented as a finite state machine, in order to keep track of the current state of the dialogue and ensures that interactions are consistent with respect to the dialogue structure.

The **message manager** processes the content of messages received from and sent to other agents. Regarding the incoming messages, it is responsible for the semantic interpretation of their content, translating it into an RDF representation that can be queried as a knowledge base by the decision maker through the communication manager (e.g. the Decision Manager can be asked if any agent has sent a notification about a payment being done). For outgoing messages, the message manager is responsible for the conversion of the agent's internal representation into the common, understandable format, used by all the agents. Translation rules can be defined according to ontological relations.

3.3   Supporting Systems

At least two supporting system components are needed by contract-aware agents: a *contract repository* and an *ontology store*. The Contract Repository provides persistent storage of *contract templates* and *contract instances* avoiding the loss of information about contracts between sessions. The Ontology Store is a repository for storing and retrieving domain and contractual knowledge, which provide ontological definition of terms, predicates and actions which are referred to in the terms of the contracting language. These components can be deployed either only in the contracting environment or inside each contract-aware agent extending the core set of components previously described.

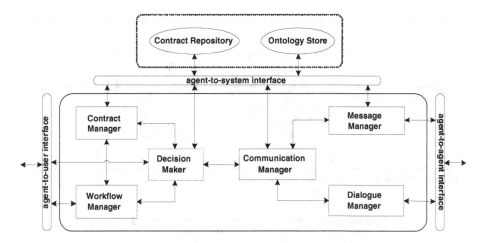

Fig. 2. Agent Service Middleware Architecture

4 Example of Contract Management and Execution

In this section we describe how business and administrative contract parties are used in the design of a simple contract-based application, an on-line music store. We further illustrate how the internal modules we propose can be used. The scenario is depicted in Figure 3. Business agents, such as the *musicBuyer* and *musicSeller*, are the signatory parties of the *buyMusicContract* and are involved in contractual interactions. The *Notary* and the *Conflict Manager* are administrative contract parties which respectively support the establishment and monitoring of the *buyMusicContract*. The Notary is a certification entity which supervises the contract creation between *musicBuyer* and *musicSeller*. The Conflict Manager takes the roles of manager and observer and monitors contractual interactions and is responsible to detect and solve any conflicts that can arise between the *musicBuyer* and *musicSeller* during contract execution. The Contract Repository and Ontology Store respectively provide contract storage functionalities and ontological definition to the other agents. Finally, the *Banking Service* is a third party service involved in the *musicBuyer* payment workflow. It also notifies the *musicSeller* about the outcome of the payment.

According to the contracting language defined in [6], a contract document specifies signatory parties and the deontic clauses parties are subject to. In this scenario, the *musicBuyer* and *musicSeller* agents represent the signatory parties. For this example, the *buyMusicContract* includes six clauses:

dc1. if available(DownloadService) $\longrightarrow$
  PERMITTED$_{musicBuyer}$(buyDownloadRights(CD.Id))

dc2. if done(buyDownloadRights(CD.Id)) $\longrightarrow$ OBLIGED$_{musicBuyer}$(pay(CD.Price))
  BEFORE 1 Hour

dc3. if violated(dc2) $\longrightarrow$ PERMITTED$_{musicSeller}$(abort(buyMusicContract))

dc4. if done(pay(CD.Price)) and (available(DownloadService) $\longrightarrow$
  OBLIGED$_{musicSeller}$(giveDownloadRights(CD.Id,musicBuyer)) BEFORE 1 Hours

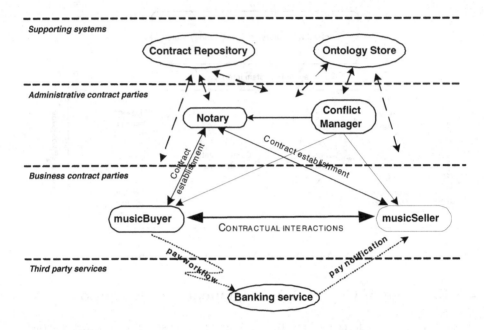

Fig. 3. The contracting environment in the *buyMusic* scenario

dc5. if done(giveDownloadRights(CD.Id,musicBuyer)) $\longrightarrow$
PERMITTED$_{musicBuyer}$(download(CD.Id) BEFORE 1 Hour)
dc6. if violated(dc5) $\longrightarrow$ OBLIGED$_{musicSeller}$(payDelivery(Delivery.fee) and (deliverCD(CD.Id) BEFORE 7 Days))

Clauses are expressed through a variation of the representation defined in [18], which is based on a dyadic deontic logic including conditional and temporal aspects. Clauses are associated with a well-defined lifecycle. When a contract is agreed, clauses are *inactive*; they move into an *active* state when their associated activating conditions hold. Then, they may change to a *fulfilled* state when their associated actions are executed or to a *violation* state when their associated actions fail. Clause state changes are perceived through the message manager interpretation process of agents' messages. These percepts are forwarded to the contract manager (through the communication and decision maker modules), who changes the state of the clauses accordingly.

The run-time management of contract-bounded interactions goes as follows. A new instance of the *buyMusicContract* is created from a contract template whenever the *musicBuyer* starts the process of buying music on the online *musicStore*, managed by the *musicSeller*. At this stage, the Communication Manager, the Dialog Manager (and Message Manager) are involved in the contract establishment process. The Communication Manager provides the *musicBuyer* with details such as the IP address and port number for contacting the *musicSeller* and the protocol to be adopted (or the opposite, depending on which one of the two is the initiator). A Simple Contract Create Protocol can be used (see [6] for protocol's details) to negotiate and instantiate a new contract and the Dialogue Manager makes sure that the protocol is followed correctly.

Once the *buyMusicContract* has been settled, the Decision Maker module extracts the deontic clauses, the agent is responsible for. This knowledge about the status of the contract and its clauses, is kept by the Contract Manager. Agents are now ready to execute the signed contract. It is important to clarify here that clauses of a contract implicitly define a (partial) ordering in the sequence of the actions to be performed when executing the contract, i.e. the workflow (of the process described in the contract). The derived workflow of the *buyMusicContract* is shown in Figure 4.

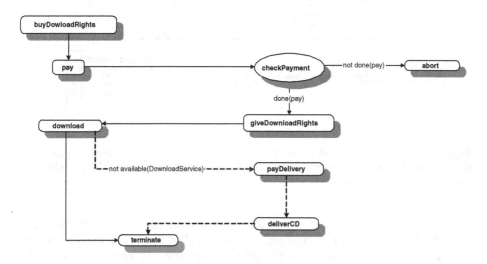

Fig. 4. The workflow derived from the *buyMusicContract* clauses. During the *download* an exception may occur which triggers the activation of violation clause dc5 and the dotted path of the workflow is followed.

The first action to be executed is the *buyDownloadRights*. According to clause dc1, download rights can be bought if the service is available. The Communication Manager converts the *musicBuyer*'s intention to buy in a communication goal in terms of a request protocol (Figure 5(a)). The translation of the *musicBuyer* action representation into a understandable representation for the *musicSeller* is done by the Message Manager. In this case, the *buyDownloadRights* action is translated in a REQUEST(*sellDownloadRights(CD.Id)*) message. The request protocol is started by the Dialogue Manager. When an agree messages is received, the Workflow Manager will be responsible of carrying on the agreed action. The reception of an INFORM-DONE message will trigger the fulfillment of clause dc1 and the activation of clause dc2 as its activating condition holds. The *musicBuyer* and *musicSeller* interactions for the buy action and their clause states are shown in Figure 5(a).

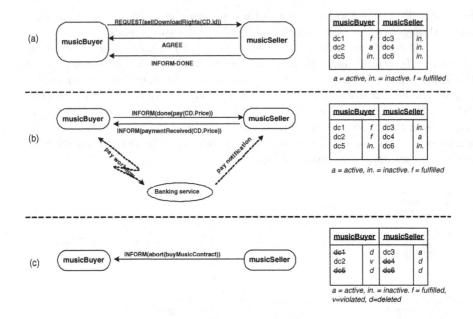

Fig. 5. The *musicBuyer* and *musicSeller* interactions and internal states regarding the contract clauses for the *buyDownloadRights* action (a), *pay* action (b) and the *pay* clause violation (c) respectively

Next, the buyer is obliged to pay for the download. To accomplish that, the *musicBuyer* interacts with a banking service according to a pre-defined payment workflow. Once finished, he acknowledges the *musicSeller* by an INFORM message (again, Communication Manager, Message Manager and Dialogue Manager are involved here). When the *musicBuyer* receives the payment confirmation from the banking service as well, it informs the *musicSeller* that the payment was successfully concluded, which will update the clauses' states (Figure 5(b)). It may however happen that buyer or seller have not matching perceptions about the payment. In this case, the conflict is detected and solved by the Conflict Manager. The Conflict Manager enforces the activation of violation clause dc3, and ensures subsequent clauses become invalid and previously fulfilled clauses are deleted. The seller then is permitted to abort the contract (Figure 5(c)).

When the payment succeeds, the *musicSeller* clause dc4 becomes active and so does the obligation to give the rights for the download. When dc4 is fulfilled, dc5 becomes active and the buyer is allowed to download the CD. In the case the download service becomes unavailable because of overloading of the server during the time period clause dc5 is still active, an exception is raised and the seller is responsible to deliver the CD before 7 days and to pay the delivery fees. Whenever all contract clauses are fulfilled, the agents can start a Contract Termination Protocol (see [6] for protocol's details). When the protocol ends, the contract can be considered successfully terminated.

5   Related Work

Contracts and Service Level Agreements are emerging as an important area in Service oriented systems. As a result there are quite a number of initiatives underway. One of the best known is Cremona [8] – a SLA middleware function complementing the basic Web Services stack. Cremona helps providers to read, fill in and manage agreement templates, implement the agreement protocol, check availability of service capacity and monitor agreement states at runtime. Cremona's communication component handles various message types and message sequencing to form interaction protocols through the agent knowledge bases. Though, the framework seems not be competent on decision making based on the content of an agreement, job scheduling and resource management. Also the monitoring components in Cremona do not provide support for agreement breaking. Finally, workflow handling, which is essential when dealing with complex interactions between services, is not supported.

The Web Service Level Agreement (WSLA) [3] is a framework targeted at defining and monitoring SLAs for Web Services [9]. The general structure of an SLA in WSLA includes the involved parties, the SLA parameters, the metrics and algorithms to compute those parameters, the service level objectives (SLOs) and the actions to be taken if a violation has been detected. The WSLA Framework implementation is based on the IBM Web Services Toolkit and licensed as commercial software. Its main functionalities are the definition, negotiation, deployment, monitoring, and enforcement of SLAs.

However, WSLA does not fully support multiple consumer/provider contracts, as these are signed by only two parties. Concerning partner responsibilities, it should be noted that defining them exclusively in terms of parameters and metrics is quite limiting, and there is no support for the definition of actions to be fulfilled. Finally, another very important issue is the lack of a generic, flexible and automatically executable mechanism for corrective management actions.

In the case of WS-Agreement [10], the general structure of agreements consists of the description of the context in which the agreement is established, the service itself and the guarantee terms. The WS-Agreement specification is less focused on the description of the related activities that should be choreographed but on the definition of the commitments and penalties.[4] WS-Agreement is quite often used with the conversation definition language WSCL [11]. The WSCL/WS-Agreement over IBM's ETTK is the reference (but partial) implementation of these specifications. It is built on top of Cremona and extends it by using WS-Agreement templates, richer message types, and XML-based interaction protocols. However, a WS-Agreement-based framework has quite some limitations: it does not include the specification of third parties working in the management of contracts, and no metrics are defined in order to support flexible monitoring implementations over the Web service choreographies.

---

[4] WS-Agreement is the only specification of those under study that includes the explicit declaration of *penalties*, but they consist only of sets of actions.

PANDA [12] is a project developing technology for the negotiation, monitoring and evaluation of contracts in supply-chains of producers in modular distributed Enterprise Resource Planning (ERP) systems. The infrastructure combines centralized Web service-based components (catalogue of partner profiles, SLA templates storage, etc.) with distributed peer-to-peer components implemented using JADE multi-agent platform. PANDA is an attempt to connect, at a high level, Service Level Agreements and Multi-Agent Systems, focusing on semi-automated negotiation and offline monitoring of contracts, and including interesting features such as matchmaking, negotiation, and Virtual Organisation (VO) evaluation. The main issues concerning PANDA for its use in a generic contracting architecture are that its main focus is in the domain of ERP-solutions, and that it is still under development.

Recently many event-driven (ECAs) Web standards have occurred, with particular emphasis on reactive RuleML languages and their SLAs-tailored variant, namely RBSLA (Rule-based Service Level Agreements) [13]. Though, RBSLA does not explicitly adopt the Web service-technology. Standard generic rule and inference engines such as Mandarax [14], based on RuleML [15] and Prova [16] have been developed to execute and manage contracts designed in RBSLA. Prova in specific supports complex reaction rule-based workflows, rule-based complex event processing, distributed inference services, rule interchange, rule-based decision logic and dynamic access to external data sources, web-based services and Java APIs. Nevertheless, both frameworks have an object oriented aspect, concentrate on the support of inference and reasoning engines, do not provide direct support for contracting procedures between agents and operate on a declarative rather than deontic level.

Each of these existing pieces of work provide a different perspective on contracts and Web services, however tackle the problem from a particular perspective such as language (RBSLA), negotiation (Cremona) or specification of the contract content (WSLA). The aim of the work presented here is to provide an implementation which matches a formal model of an overall contracting process and environment in a more general way [4].

6 Conclusions and Future Work

In this paper we have presented a middleware architecture to support contract-aware agent-based services development. The middleware consists of several components, including a *contract manager*, a *communication manager* and a *workflow manager*, which together help the main reasoning cycle of the agent (the *decision maker* module) to cope with the low-level management of such aspects.

The proposed middleware is built on top of a contracting environment [7] which is a Web service implementation of the IST-CONTRACT framework [4]. The result is that designers can build agent-oriented Web services which can 1) create contracts, 2) handle all contract-related communication, 3) manage the active responsibilities during the contract execution and 4) solve disputes if third, administrative parties are included in the system. As internal components communicate through Java APIs, designers can decide to plug-in their own developed modules (such as a workflow manager).

We have shown through the description of a simple scenario how the internal components of the middleware can be used. Compared with other SLA technologies, designers modeling effort slightly increases, by having to describe the agreements not in terms of metrics but on higher level abstractions (such as obligations, permissions, prohibitions, violations) and providing sound action descriptions and ontologies. But the increase of expressiveness achieved compensates the increase of modeling effort, extending the use of electronic contracting to more complex, dynamic settlements. Moreover the formal specification of contracts through deontic clauses (grounded with clear defined semantics) and the specification of actions in terms of pre- and post-condition, allow the development and use of off-line and runtime verification techniques of desired properties of the system through, e.g. model-checking techniques [19] (work which is actually being carried on in the context of the IST-CONTRACT project).

The middleware is now being tested in three case scenarios in the context of the CONTRACT project. Current version uses a simple reasoning cycle in the Decision Maker module implemented in 2APL. One of the next steps is to try to adapt other intelligent agents technologies, such as Drools or JADEx. Another planned improvement is to implement a subscribing mechanism between modules that permits, e.g. that the Contract Manager receives updates directly from the Communication Manager whenever things in the interaction are related to actions or predicates that can change the state of a clause.

Future plans include the addition of a wider variety of protocols and formats (including where possible compatibility with frameworks such as WS-Agreement) and deployment on robust platforms and open source Web service containers.

Acknowledgments

This work has been funded mainly by the European Commission Framework 6 funded project CONTRACT (INFSO-IST-034418). Javier Vázquez-Salceda's work has been also partially funded by the Ramón y Cajal program of the Spanish Ministry of Education and Science. All the authors would like to thank the CONTRACT project partners for their inputs to this work.

References

1. OASIS ebXML Joint Committee: ebXML Website (2008), http://www.ebxml.org
2. Verma, D.: Supporting Service Level Agreements on IP Networks. Macmillan Technical Publishing (1999)
3. Ludwig, H., Keller, A., Dan, A., King, R.P., Franck, R.: Web Service Level Agreement (WSLA) Language Specification (2003), http://www.research.ibm.com/wsla/WSLASpecV1-20030128.pdf

4. Miles, S., Oren, N., Kollingbaum, M., Luck, M., Álvarez-Napago, S., Vázquez-Salceda, J.: Contract based Electronic Business Systems Theoretical Framework. IST-CONTRACT project deliverable D2.2 (October 2007), http://www.ist-contract.org
5. Wooldridge, M.: Introduction to Multi-agent Systems. John Wiley and Sons, Chichester (2002)
6. Panagiotidi, S., et al.: Contract Language Syntax and Semantics Specifications. IST-CONTRACT project deliverable D3.1 (October 2007), http://www.ist-contract.org
7. Biba, J., Confalonieri, R., Willmott, S., Jakob, M., Dehn, M., Bangel, D., Álvarez-Napagao, S.: Web Services Framework for Contract Based Computing. IST-CONTRACT project deliverable D4.1 (October 2007), http://www.ist-contract.org
8. Ludwig, H., Dan, A., Kearney, R.: Cremona: an architecture and library for creation and monitoring of WS-agreements. In: Proceedings of 2nd International Conference on Service Oriented Computing 2004, pp. 65–74. ACM, New York (2004)
9. Keller, A., Ludwig, H.: The WSLA Framework: Specifying and Monitoring Service Level Agreements for Web Services. Journal of Network and Systems Management 11(1), 57–81 (2003)
10. Andrieux, A., et al.: Web Services Agreement (WS-Agreement) Specification (2005), http://www.globalgridforum.com/Public_Comment_Docs/Documents/Oct-2005/WS-AgreementSpecificationDraft050920.pdf
11. Banerji, A., et al.: Web Services Conversation Language (WSCL) 1.0. W3C Note (2002), http://www.w3.org/TR/wscl10/
12. PANDA: Collaborative Process Automation Support using Service Level Agreements and Intelligent dynamic Agents in SME clusters. IST-PANDA research project (2007), http://www.panda-project.com/
13. Paschke, A.: RBSLA: A declarative Rule-based Service Level Agreement Language based in RuleML. In: Proceedings of the International Conference on Computational Intelligence for Modelling, Control and Automation and International Conference on Intelligent Agents. Web Technologies and Internet Commerce, vol. 2, pp. 308–314 (2005)
14. Paschke, A.: The Mandarax RDF / RDFS/ OWL / DLP Module - Integration of Semantic Web Data into the Rule Engine Manadarax. In: International Workshop on Rule-Based Modeling and Simulation of Interacting Systems and Agents (AORML), Cottbus, Germany (February 2006)
15. Boley, H.: The rule-ml family of web rule languages. In: 4th Int. Workshop on Principles and Practice of Semantic Web Reasoning, Budva, Montenegro (2006)
16. Kozlenkov, A., Schroeder, M.: PROVA: Rule-based Java-Scripting for a Bioinformatics Semantic Web. In: Rahm, E. (ed.) DILS 2004. LNCS (LNBI), vol. 2994, pp. 17–30. Springer, Heidelberg (2004)
17. Panagiotidi, S., Vázquez-Salceda, J., Álvarez-Napagao, S., Ortega-Martorell, S., Willmott, S., Confalonieri, R., Storms, P.: Intelligent Contracting Agents Language. In: Workshop on Behaviour Regulation in Multi-Agent Systems (BRMAS 2008) (April 2008) (accepted)
18. Vázquez-Salceda, J., Aldewereld, H., Dignum, F.: Implementing norms in multi-agent systems. In: Lindemann, G., Denzinger, J., Timm, I.J., Unland, R. (eds.) MATES 2004. LNCS (LNAI), vol. 3187, pp. 313–327. Springer, Heidelberg (2004)
19. Lomuscio, A., Qu, H., Solanki, M.: Towards verifying compliance in agent-based web service compositions. In: Proceedings of the 7th International Conference on AAMAS, May 2008. Estoril, Portugal, IFMAS Press (2008)

A Knowledge Technologies-Based Multi-agent System for eGovernment Environments

Francisco García-Sánchez[1], Luis Alvarez Sabucedo[2], Rodrigo Martínez-Béjar[1], Luis Anido Rifón[2], Rafael Valencia-García[1], and Juan M. Gómez[3]

[1] Univeridad de Murcia, Spain
{frgarcia,rodrigo,valencia}@um.es
[2] Universidade de Vigo, Spain
{lsabucedo,lanido}@det.uvigo.es
[3] Universidad Carlos III de Madrid, Spain
juanmiguel.gomez@uc3m.es

Abstract. The increasing volume of eGovernment-related services is demanding new approaches for service integration and interoperability in this domain. Semantic Web technologies and applications can leverage the potential of eGovernment service integration and discovery, facing the problems of semantic heterogeneity of eGovernment information sources and the different levels of interoperability. In line with this, eGovernment services will be semantically described in the foreseeable future. In an environment with semantically-annotated services, software agents are essential as the entities responsible for exploiting the semantic content in order to automate some tasks thus improving the user experience.

In this paper, we first present SEMMAS, an ontology-based Multi-Agent framework for seamlessly integrating Intelligent Agents and Semantic Web Services. The proposed framework is independent from both the application and the domain. Our approach is backed with a proof-of-concept implementation where the breakthrough of integrating disparate eGovernment services has been tested.

1 Introduction

Agent Technology has been broadly studied over the last 30 years and is currently being revisited due to its relation to the Semantic Web (SW) and the potential benefits that can be reached from their integration. An Intelligent Agent (IA) can be defined as a computer system situated in some environment and capable of autonomous action in this environment in order to meet its design objectives [21]. An IA is characterized by a set of basic properties including reactivity, proactiveness, and social ability. Before the emergence of the SW, agents had to face the problems derived from the lack of structure in the information published on the Web. The SW [1] involves the addition of machine-readable, semantic annotations to Web resources by using ontologies [17] as the backbone technology. Hence, in the so called Web 3.0, agents will be able to automatically process and exploit the machine-readable, semantic contents of the Web and new powerful opportunities will open up for both application developers and users.

R. Kowalczyk et al. (Eds.): SOCASE 2008, LNCS 5006, pp. 15–30, 2008.

On the other hand, Web Service Technology have arisen as the best solution for remote execution of functionality. This is partly due to properties such as operating system and programming language-independence, interoperability, ubiquity and the possibility to develop loosely-coupled systems. Web Services (WSs) aim to transform the Web from a mere repository of information into a distributed source of functionality. However, as the Web grows in both size and diversity, there is an increased need to automate aspects of WSs such as discovery, execution, selection, composition and interoperation [3]. Semantic Web Service (SWS) technology, that is, the semantic annotation of services capabilities, is the proposed solution. The W3C is currently examining various approaches with the purpose of reaching a standard for this technology including OWL-S[1], WSMO[2], SWSF[3], WSDL-S[4], and SAWSDL[5].

Both IA and SWS technologies are able to reach remarkable achievements and in some cases have overlapping functionalities. However, independently of the undoubted benefits of these technologies, they suffer from different problems that limit their functionality when applied separately and prevent them from being implanted at a massive scale in industry [12,13]. Thus, though the Intelligent Agents and Web Services paradigms are often viewed as similar and competing, several research studies have demonstrated that the cooperative interaction between them can lead to the development of new more powerful applications [4,5,16]. In this paper, we present SEMMAS, a domain-independent framework that succesfully integrates SWSs and IAs. SEMMAS is based on a loosely-coupled infrastructure and makes use of ontologies to facilitate agents and services interoperation. With this approach, applications can benefit from the autonomy, pro-activeness, dynamism and goal-oriented behaviour IAs provide, and the high degree of interoperability across platforms WS technology advocate.

At the same time, a huge effort in eGovernment development is taking place nowadays. The provision of eGovernment solutions involves a tendency in all public services-related processes, by putting the citizen in the centre of the process. In line with this, several research studies (e.g. [6,7,9]) have identified eGovernment as an appropriate test-bed for effectively evaluating technologies such as SWSs in real-world settings. Following that trend, the SEMMAS framework has been applied to a use case scenario in eGovernment.

The rest of this paper is organized as follows. In Section 2, various tools and research studies concerning the integration of IAs and SWSs are analyzed and the current trend in the eGovernment domain briefly described. The ontology-based framework for joining together agents and services is formulated in Section 3. In Section 4, the proposed framework is applied to an eGovernment use case. Finally, conclusions and future work are put forward in Section 5.

---

[1] Web Ontology Language for Services, http://www.w3.org/Submission/OWL-S/
[2] Web Service Modeling Ontology, http://www.w3.org/Submission/WSMO/
[3] Semantic Web Services Framework, http://www.w3.org/Submission/SWSF/
[4] Web Service Semantics, http://www.w3.org/Submission/WSDL-S/
[5] Semantic Annotations for WSDL, W3C Recommendation from August 28th, 2007; http://www.w3.org/TR/sawsdl/

2   Related Work

2.1   Multi-agent Systems and Semantic Web Services

The problem of how the ontology languages of the SW could lead to more powerful agent-based approaches to using services offered on the Web was first described by Hendler [10]. In his work, the author presented the foundations of what is now called SWS. He proposed a method for describing the way the invocation of services should be done by agents by means of an ontology language such as DAML+OIL. Once the invocation characteristics of a service are semantically described, agents would be able to determine the specific information needed for invoking the service.

More recently, there have been various research projects that have investigated methods to support the cooperative interaction between SWS and IA. Blacoe and Portabella [2] point out that, in order to integrate what they called 'agent-based services' and 'web-based services', three main solutions emerge. WS can provide the most basic level functionality while agents can supply higher-level functions by using, combining and choreographing WS, so achieving added-value functions. Alternatively, communication in WS and agents may become equivalent, so that there is no distinction between them ('agents in web service wrappers'). Finally, both concepts can remain separate creating a heterogeneous service space and interoperating through gateways and translation processes.

The Semantic Web FRED project (SWF) combines agent technology and ontologies in order to develop a system for automated cooperation [16]. In this system, software agents (called 'Freds') perform tasks on behalf of their owners and interact among them if they have to. In order to resolve a task, agents make use of computational resources that allows automated resolution of tasks, referred to as 'services'. The authors distinguish three types of services: plans (i.e. Java programs), processes (i.e. complex and nested services), and external WS (through their WSDL interface). A major problem of the SWF is that is tightly bound to WSMO. In fact, the SWF is supposed to be a "WSMO Implementation" unlike the framework presented here, which includes mechanisms to support all the current SWS approaches. Besides, with our approach it is even possible to dynamically incorporate support for new solutions to the semantic annotation of services.

The GODO (Goal Oriented DiscOvery) system [4] can be described as an agent located between the users and the WSMX environment. When users wish to send goals to WSMX, they have to write them down in WSML, a formal language that can be hard to understand for non-expert users. GODO is able to transform user requests in natural language into WSMX goal in WSML. With this purpose, it incorporates a language analyzer that determines the ontology-relevant elements contained in each sentence, thus producing a lightweight ontology. GODO uses then the ontology to generate the goals to be executed and sends them to WSMX. GODO's main problem is that it interacts with an execution environment and not with SWS as such. Therefore, not all the benefits of using agent systems are exploited.

The "Agents and Web Services Interoperability Working Group (AWSI WG)" aims to create a middleware able to handle the fundamental differences between Agent Technology and Web Services [5]. Aligned with this approach is the AgentWeb Gateway middleware [15], which facilitates the required integration without changing existing specifications and implementations of both technologies. This solution complies with the third category within Blacoe and Portabella's classification. However, in our opinion it is not conceptually appropriate that WS and IA work at the same (abstraction) level, not to mention the inconveniency of ignoring the fundamental differences between the WS and IA paradigms.

2.2   ICTs in eGovernment

We can find several definitions of eGovernment from relevant sources [18,19]. Nevertheless, all of them are mainly referred to the provision of advanced services in the domain of Public Administrations (PAs) for a better service from the point of view of the citizen taking advantage of ICTs. In order to develop solutions in the domain, entities and organizations from all over the world are devoting a large amount of resources to this area. There are two main reasons for this phenomenon. On the one hand, citizens are demanding a more efficient and friendly administration and, on the other hand, Governments worldwide are creating laws to empower this sort of technologies at all levels. Spain is a clear expample of this new trend. The Spanish parliament recently approved a law[6] to guarantee the access to services provided by public administrations under the support of ICTs.

Among national initiatives, those supported in an official manner by countries, we can mention SAGA[7] in Germany, ADEA[8] in France, e-GIF[9] in the United Kingdom, or FEAF[10] in USA. The analysis of these projects [8] may lead to establish some conclusions about them. At first glance, it is clear the lack of real indications, a solid framework, a data model or even software infrastructure to address the problem of providing solutions in a eGovernment environment. Most of the presented proposals just provide some general bases or recommendations for the development of software products under some generalist ideas.

Besides those national projects, we can outline some projects founded by different entities, mainly the European Union, to fulfill solutions in this area:

OntoGov: this project addresses the problem of services in eGovernment under a semantic point of view and it is aimed to provide an ontology to model the problem in a maintainable way[11].

Terregov: this project's main goal is to provide an interoperable layer that allows citizens to access eHealth services in a transparent manner by means of web services[12].

---

[6] http://www.boe.es/boe/dias/2007/06/23/pdfs/A27150-27166.pdf

[7] http://www.kbst.bund.de/saga

[8] http://www.adae.gouv.fr/adele/

[9] http://www.govtalk.gov.uk/

[10] http://government.popkin.com/frameworks/feaf.htm

[11] http://www.ontogov.com/

[12] http://www.terregov.eupm.net/my_spip/index.php

EPRI: this project is aimed to increase the role of Information Society Technology among the administration, mainly at EU, national and regional levels[13].

The SemanticGov project: this one is aimed at developing a software infrastructure to provide support for PAs by means of semantic. Currently, it is an ongoing project[14].

EUPubli: it is a project related to the provision of a pan-European layer for intermediation.

eGovernment Good Practice Framework: the goal of this project is to recollect documentation about good practices and to make them well known.

DIP: this project is concerned with the technological background of data, information, and process integration through SWSs[15]. The results of this project have been tested in a eGovernment use case scenario [6].

Also, international organisms involved in the technological development are playing a role in this area:

DGRC. The Digital Government Research Center was founded in 1999 by the NSF (National Science Foundation), and its area of interest is the investigation of ICTs applied to eGovernment services. In this center, several projects have been carried out and the information to citizen is provided by means of the newsletter dgOnline.

OMG. The Object Management Group, besides all the realized projects and initiatives, launched a specific working group, namely E-Gov DSIG. Currently, this working group is at an initial stage.

OASIS. The Organisation for the Advancement of Structured Information Standards has also its own committee devoted to the study of the applicability of its own technologies to eGovernment.

CEN. The European Committee for Standardization launched its own group of interest in the area in February 2005. In addition to this, it has undertaken some interesting work by means of some CWAs (CEN Workshop Agreements): CWA 1859 *"Guidance on the Use of Metadata in E-Government"*, CWA 1860 *"Dublin Core E-Government application Profiles"* and CWA 13988 *"Guidance information for the use of Dublin Core in Europe"*.

3   The SEMMAS Framework

The approaches developed so far with the aim of integrating IAs and SWSs suffer from shortcomings mainly due to their inability to overcome the problems associated to each of the technologies under question and completely benefit from the advantages of their combination. The framework presented here stems from a basic underlying hypothesis: IAs and WSs must lie on two different layers of abstraction due to the conceptual differences between these technologies

---

[13] http://www.epri.org/
[14] http://www.semantic-gov.org
[15] http://dip.semanticweb.org/

from their very conception. Certainly, the main idea behind Agent Technology was not for IAs to be able to provide services, but to become autonomous entities that incorporate intelligence, being capable of exhibiting pro-active, goal-oriented behaviour, and interacting (either competitively or cooperatively) with other entities in order to satisfy their design objectives. In contrast, WSs were conceived with the purpose of providing globally accessible software components that expose particular functionalities. It is essentially the same idea as with roles in any job: depending on your background and profile you will be better prepared for one kind of task or another. Similarly, why using IAs to carry out (repetitive, non-knowledge intensive) tasks that can be performed by WSs, if IAs are more appropriate for exploiting the functionality provided by those WSs and improving the user experience.

Next the foundations of the referred framework are presented and the main elements of the architecture enumerated.

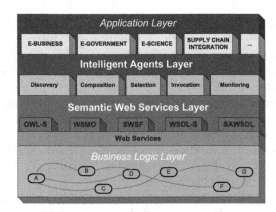

Fig. 1. Multi-layered infrastructure

3.1 Fundamentals

Ontologies are the paramount technology of the approach introduced in this paper as they act as the 'glue' that binds together the remainder components of the framework. In this work, an ontology is seen as "a formal and explicit specification of a shared conceptualisation" [17]. Firstly, ontologies function as domain vocabularies so that Web Services and agents share the same interpretation of the terms contained in the messages that they exchange. Secondly, ontologies are useful to semantically describe Web Services capabilities and processes. This semantic description can then be automatically processed by software entities, so that Web Service discovery, composition, selection, execution and monitoring can be done without human intervention. Finally, from the agents' perspective, each agent's local domain-related knowledge may be extracted from, or built upon, the application domain ontology. Moreover, inter-agent communication may be carried out by means of a common vocabulary based on an agreed ontology.

The framework presented here is based on a multi-tier architecture that is composed of four different layers (see Fig. 1). The lower layer, namely, the Business Logic Layer provides for the most specific operations. It comprises the internal business processes within companies. It usually consists of legacy systems. Upon this layer, WSs are deployed that show off some parts of the internal business process and make that functionality publicly available. These services along with the semantic description of their capabilities lie on the second layer, namely the Semantic Web Services Layer. Adding semantic annotations to WS capabilities can help software entities to (semi-)automatically interact with them in a dynamic way. In particular, new services can emerge and others may change their functionality or even disappear at run-time, but the system would keep on working and the changes would be reflected on the application instantly. These sophisticated software entities (i.e. IAs) that interact with, and take advantage of, basic services are located in the Intelligent Agents Layer. IAs make use of the semantic annotation of services capabilities to automatically discover, compose, invoke and monitor WSs. They are also able to dynamically exhibit and propagate the changing functionality provided in lower layers. Finally, the Application Layer is responsible for organising (i.e. orchestrating and coordinating) agents to actually perform useful activities for users. In this way, depending on the agents available in the system and the way they inter-operate, different user-tailored applications can be obtained.

3.2 Architecture

The SEMMAS (SEMantic web service and Multi-Agent System) framework comprises two of the layers identified above, the Intelligent Agents and the Semantic Web Services layers. As a result, the framework becomes independent of both the application domain and the actual applications to be developed. In order to create an application, programmers only have to set the appropriate domain ontologies and decide on which agents to instantiate and which services to access. Thus, the framework can be considered as a reference architecture for several scenarios and complex environments such as e-commerce, e-science or e-government.

Multi-Agent System. The architecture that constitutes SEMMAS is composed of three main components (see Fig. 2): a set of IAs that constitute a MAS, four ontology repositories, and three different interfaces for interacting with the external actors that have been identified (i.e., service providers, service requesters and software developers).

In the platform proposed to run the system, three main groups of agents are distinguished: agents that act on behalf of service owners ('Provider Agent' and 'Service Agent'), agents that act on behalf of service consumers ('Customer Agent', 'Discovery Agent', and 'Selection Agent'), and agents that perform management tasks ('Framework Agent' and 'Broker Agent'). Those acting on behalf of service owners manage the access to services and ensure that the contracts are fulfilled. On the other side, the agents that act on behalf of service consumers have to locate services, agree on contracts, and receive and present results. Management agents have a double function: to balance the system workload, and to

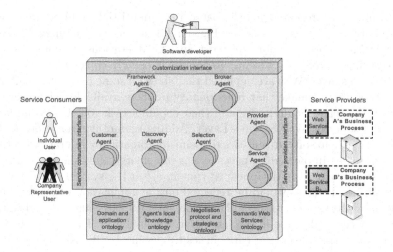

Fig. 2. The SEMMAS Architecture

help in solving the interoperability mismatches. The behaviour an agent shows at run-time depends on the goal it pursues at that time. For this purpose, the 'role' concept was introduced. Roles are encapsulations of dynamic behaviour and properties that can be played by agents. The use of roles presents a number of benefits that can be summarized as follows [22]: (1) roles are dynamic and flexible; (2) roles are responsibility-driven; (3) roles are context-sensitive. We distinguish between roles dealing with service-related issues from those related to the framework management. The service-related roles are as follows:

1. *Broker role:* it represents the functionality needed for solving all kind of interoperability problems (data, process and functional mediation).
2. *Composer role:* it allows the achievement of a goal by means of several composed services.
3. *Invoker role:* it invokes a Web Service once the operation to be executed and the parameters are known.
4. *Matchmaker role:* it finds the services whose semantic descriptions match the goal that was sent by the user.
5. *Monitor role:* it ensures that the contracts established for the execution of the services by both service owners and service consumers are fulfilled.
6. *Ontology Manager role:* it includes functionality associated with the access and processing of ontologies.
7. *Selector role:* it provides the functionality necessary for the selection of a service from a list of services according to a set of preferences.

The platform management roles are described next:

1. *Negotiator role:* it enacts the actual negotiation process between the parties establishing the basis for the system execution.

2. *Platform manager role:* it incorporates functions to control and balance the system workload.
3. *Provider Representative role:* it interacts with service providers. At a high level of abstraction, it must be able to enforce the conditions present in the company's business strategy.
4. *Service Representative role:* it acts on behalf of services, participating in negotiations and improving the offered services when possible.
5. *Consumer Representative role:* it interacts with service consumers by, firstly, determining their wishes and, then, returning the expected results.
6. *Global Monitor role*: it monitors the events in the application, detects possible problems, and defines the actions to take in case of error.

At run-time each agent decides on what roles to play depending on the goal it pursues. Nevertheless, some of the roles are mandatory for some agents, as they characterize the type of agent the agents belong to. Thus, for example, the framework agents should take over both the 'Global Monitor' and the 'Platform Manager' roles. Both factors, the actual implementation of these roles and the agent election of the roles to take, eventually determine the agent behaviour.

Ontology Repositories. In order for IAs to successfully carry out their assigned tasks, they must have access to various data repositories containing the knowledge that is necessary to fulfil the assignment. These repositories can be either local or external to the system. Knowledge in the framework is represented by means of ontologies. Four kinds of ontologies have been identified:

Application and domain ontology: the application ontology contains the knowledge entities (i.e. concepts, attributes, relationships, and axioms) that model the application in which the framework is to be employed. On the other hand, the domain ontology represents a conceptualization of the specific domain the framework is going to be applied in. This ontology supports the communication among the components in the framework without misinterpretations.

Agent local knowledge ontology: it contains, for each agent, the knowledge about the environment it possesses. This ontology generally includes knowledge about the assigned tasks, as well as the mechanisms and resources available to achieve those tasks. Thus, for example, the Broker Agent's local knowledge ontology may contain the mapping rules it has to apply to resolve the interoperability mismatches that might occur during the system execution.

Negotiation ontology: it comprises both negotiation protocols and negotiation strategies that constitute the negotiation mechanisms agents must use to coordinate their interactions. With this ontology, agents can choose the best mechanism to use for coordinating their actions, which highly depends on the problem under question and the application domain.

Semantic Web Services ontologies: in this repository (that can be comprised of various ontologies distributed all over the Internet) the ontologies that contain the semantic description of Web Services are stored. The framework does not impose any restriction in terms of the kind of SWS specification (i.e., OWL-S, WSMO, SWSF, WSDL-S or SAWSDL) to be used.

User Interfaces. At last, three different interfaces have been included within the framework architecture. They aim at enabling the interaction with the actors that are external with respect to the framework: service consumers, service providers, and software developers. Software developers can, by means of their interface, customize the application by setting up the specific ontologies to be used. They also have to instantiate and configure the core agents necessary for the proper functioning of the system (customer, provider and service agents will be launched as needed at run-time). Once the application has been properly set up, both service consumers and service providers can register in the system and use it as a meeting point. Through their interface, service providers can modify the list of services they provide and set the conditions under which a service they provide must be executed. Service consumers, on the other hand, can, by means of their interface, query the system and trigger the execution of one or several Web Services in order to fulfil a particular goal.

4   SEMMAS in eGovernment

Based on the above presented framework, a MAS was developed. For the design of this complex MAS the INGENIAS methodology [14] was applied. One of the main reasons for choosing INGENIAS was that, while most of the other methodologies do not have any reference implementation for developing the models and diagrams proposed by the methodology, INGENIAS comes along with a toolkit, the INGENIAS Development Kit (IDK)[16]. It provides the means to create most of the diagrams and models required by the methodology making it more convenient for the modeller to design the system. Besides, it incorporates a function to automatically generate JADE files from the developed diagrams. JADE (Java Agent Development Framework)[17] is the agent platform that has been used for implementing the system. JADE is the most widely used agent platform for research projects worldwide and it seems to be at a quite mature stage of development already (JADE version 3.4 has been used for the implementation of the framework).

Apart from JADE, various tools and libraries has been used to develop the framework. The latest Web programming techniques based on JavaServer Pages (JSP) and Servlets have been applied for the user interface. In order to communicate the Web interface with the MAS, the JadeGateway[18] was employed. Exploiting the semantic content of ontologies has been possible through the use of the Jena API[19]. KAText[20], a natural language processing tool, has been utilised to support the interaction of non-expert users with the system. Finally,

---

[16] http://sourceforge.net/projects/ingenias/
[17] http://jade.tilab.com/
[18] http://jade.tilab.com/doc/tutorials/JadeGateway.pdf
[19] http://jena.sourceforge.net/

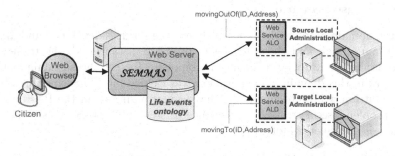

Fig. 3. eGovernment scenario

similarly to other research studies (e.g. [11]), we have followed a planning approach for service composition. In particular, we have made use of a library that implements a STRIPS-like planner[20].

4.1 Methodology

As it was stated before, SEMMAS is an application- and domain-independent framework that can be useful in various environments. In order for software developers to adapt the framework to a particular application domain they must carry out four main tasks:

1. Design the domain ontology: the scenario proposed in the next section requires a eGovernment-domain ontology that represents the knowledge concerning PAs and agencies and how they deal with changes in certain conditions. For this, we have made use of the "Change of circumstances" ontology[21], which was developed in the scope of an European project. OWL (Web Ontology Language) is the ontology language used in this work.
2. Develop the Web Services that provide the most basic functionality: two basic services ('ALO' and 'ALD') have been implemented that represent the PAs involved in the proposed scenario (see Fig. 3). Apache Axis2 was chosen as the WS implementation framework.
3. Semantically annotate the Web Services capabilities: the two abovementioned services were semantically annotated by making use of the OWL-S approach. For this, the "OWL-S Editor" Protègè plugin[22] was employed.
4. Implement the roles and instantiate the necessary agents: ad-hoc techniques for service discovery, selection, composition and invocation were developed. Service discovery is carried out by means of SPARQL queries. Utility theory methods are applied for selection while planning algorithms are employed for service composition[23] .

---

[20] http://www.dcs.shef.ac.uk/~pdg/com1080/java/strips/
[21] http://dip.semanticweb.org/documents/D9-5.doc
[22] http://owlseditor.semwebcentral.org/
[23] For more details concerning the proof-of-concept implementation visit
   http://www.semmas.com

4.2   Use Case Scenario

Once the domain ontology had been defined, the services developed and appro-
priately annotated, and the core agents properly instantiated, the framework
can be applied in an eGovernment environment. To test the suitability of the
framework in this domain, an experiment was carried out. The experiment consist of a citizen willing to change his/her regular address and the PAs involved
in the problem (see Fig. 3). The sequence of steps SEMMAS takes to achieve
that goal is described as follows:

Query analysis (see Fig. 4). At first, the system processes and interprets
the user request by translating it into an internal goal-ontology model (see
Fig. 5). For this, the 'Customer Agent' uses KAText, which is able to transform a natural-language sentence into a lightweight ontology. Users can also
input their wishes by directly setting up an ontology model with their goal.

Service discovery. Once the user goal is formally represented, the 'Customer
Agent' sends it to the 'Discovery Agent'. This agent queries the accessible
SWS repositories for services that meet the user's requirements.

Service composition. Given that none of the available services is able to
achieve the goal by itself, the composition phase of the 'Discovery Agent'

Fig. 4. Customer interface - Main menu

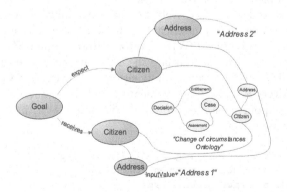

Fig. 5. User goal

starts. During this phase, the agent, by means of the planner, descomposes the goal into simpler subgoals and starts the discovery cycle again for each new subgoal. In the proposed scenario, the subgoals are "Unsubscribe old residence" and "Subscribe new residence".

Service selection. Once at least a service has been discovered for each subgoal ('ALO' for unsubscribing and 'ALD' for subscribing), the 'Selection Agent' carries out the selection stage. For this, it assigns an utility value to each candidate based on the comparison between the user's preferences and the terms the providers propose for the execution of their services during the negotiation process that takes place at this stage.

Fig. 6. Customer interface - Found services

Service invocation. In this example, only one candidate (the composition of 'ALO' and 'ALD') is present (see Fig. 6). Therefore, the system, after requesting authorization from the user, sequentially executes the selected services, collects the results ('Service Agent'), and retuns them to the user ('Customer Agent') (see Fig. 7).

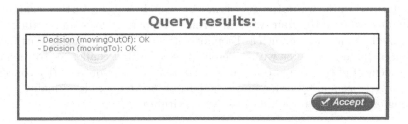

Fig. 7. Customer interface - Result

5   Conclusions

In this paper, we first introduce both Intelligent Agents and Semantic Web Services technologies and point out the potential benefits that can be reached from their combination. Then, we analyze the plausibility of integrating Web Services and Multi-Agent Systems into a combined architecture and the advantages derived from it through a comprehensive list of approaches. Based on the benefits envisioned of such integration and the drawbacks of the current solutions, we present SEMMAS, a new framework for Intelligent Agents and Semantic Web Services integration. This framework aims to exploit the striking potential of the technologies under question while overcoming their deficiencies. As opposed to other Semantic Web Services execution environments (e.g. IRS-III, WSMX), SEMMAS makes use of agent technology in such a way users are better represented within the system (e.g. their specific preferences are taken into account and personalization techniques can be applied). Besides, agents are delegated knowledge-intensive, high-level tasks for which they are more appropriate. The major advantage of SEMMAS over other integrating approaches (e.g. [4,16,15]) is that it clearly defines each technology's role. Thus, with this approach, applications can benefit from the autonomy, pro-activeness, dynamism and goal-oriented behaviour agents provide, and the high degree of interoperability across platforms Web Services advocate.

Based on the above described framework, a prototype of a Semantic Web Services execution environment has been implemented. This tool comprises a Multi-Agent System that complies with the requirements imposed by the framework, and a Web application that constitutes the user interfaces. The proof-of-concept implementation has been applied to an eGovernment-based use case scenario. This simple example permits to show how the system faces the most basic service-related issues such as discovery, composition and invocation. Through this example, the advantages of using Agent Technology for, for example, carrying out negotiation processes and engaging users in the execution process are also revealed. However, it does not fully exposes the capacities of the framework. It is worth mentioning that it is not the purpose of this work to devise innovative solutions for Semantic Web Services-related tasks (i.e. matchmaking, composition, mediation, and so on) but to provide an infrastructure capable of maximizing the outcome of the Intelligent Agents and Web Services combination.

As further work, while the eGovernment domain will be further examined, we plan to evaluate the framework in terms of its performance and usability on other application domains such as Bioinformatics and eCommerce. In particular, we aim at finding domains with different properties and challenges so that the fully potential of the framework can be analyzed. In addition, security concerns have not yet been addressed, and will be part of future upgrades. Finally, we are working on the integration of Grid services within the framework.

Acknowledgments

This work has been possible thanks to the Spanish Ministry for Science and Education through projects TIN2007-68125-C02-02, TIN2006-15140-C03-02, TIN2006-14780, TEC2006-12365-C02-01, Spanish Ministry for Industry, Tourism and Commerce through projects FIT-340000-2007-134 and FIT-340000-2007-212, the Xunta de Galicia under project PGIDIT06PXIB322285PR, and the Region of Murcia under project BIO-TEC 06/01-005.

References

1. Berners-Lee, T., Hendler, J., Lassila, O.: The semantic web. Scientific American, 34–43 (May 2001)
2. Blacoe, I., Portabella, D.: Guidelines for the integration of agent-based services and web-based services. Deliverable d2.4.4 (wp2.4), Knowledge Web project (2005)
3. Fensel, D., Bussler, C.: The web service modeling framework wsmf. Electronic Commerce Research and Applications 1(2), 113–137 (2002)
4. Gómez, J.M., Rico-Almodóvar, M., García-Sánchez, F., Toma, I., Han, S.: Godo: Goal oriented discovery for semantic web services. In: Discovery on the WWW Workshop (SDISCO 2006), Beijin, China (2006)
5. Greenwood, D., Lyell, M., Mallya, A.U., Suguri, H.: The ieee fipa approach to integrating software agents and web services. In: Proc. of the 6th Int. Joint Conf. on Autonomous Agents and Multiagent Systems (AAMAS 2007), Honolulu, Hawai'i, p. 276 (2007)
6. Gugliotta, A., Domingue, J., Cabral, L., Tanasescu, V., Galizia, S., Davies, R., Gutierrez-Villarias, L., Rowlatt, M., Richardson, M., Stincic, S.: Deploying semantic web services-based applications in the e-government domain. Journal on Data Semantics X. Lecture Notes in Computer Science, vol. 4900, pp. 96–132 (2008)
7. Gugliotta, A., Tanasescu, V., Domingue, J., Davies, R., Gutiérrez-Villarías, L., Rowlatt, M., Richardson, M., Stincic, S.: Benefits and challenges of applying semantic web services in the e-government domain. In: Semantics 2006, Vienna, Austria (2006)
8. Guijarro, L.: Analysis of the interoperability frameworks in e-government initiatives. In: Traunmüller, R. (ed.) EGOV 2004. LNCS, vol. 3183, pp. 36–39. Springer, Heidelberg (2004)
9. Gutiérrez-Villarías, L., Davies, R., Rowlatt, M.: Developing sws for e-government. In: Proc. of the 3rd European Semantic Web Conference, Budva, Montenegro (2006)
10. Hendler, J.: Agents and the semantic web. IEEE Intelligent Systems 16(2), 30–37 (2001)
11. Kuter, U., Sirin, E., Parsia, B., Nau, D., Hendler, J.: Information gathering during planning for web service composition. Journal of Web Semantics 3(2-3), 183–205 (2003)
12. Louis, V., Martinez, T.: The jade semantic agent: Towards agent communication oriented middleware. AgentLink News 18, 16–18 (2005)
13. Paolucci, M., Sycara, K.: Autonomous semantic web services. IEEE Internet Computing 7(5), 34–41 (2003)
14. Pavón, J., Gómez-Sanz, J., Fuentes, R.: The INGENIAS Methodology and Tools. In: Giorgini, B.H.-S.P. (ed.) Agent-Oriented Methodologies, pp. 236–276. Idea Group Publishing, USA (2005)
15. Shafiq, O., Suguri, H., Ali, A., Fensel, D.: A first step towards enabling interoperability between software agents and semantic web services: Multi agent systems adapting web services standards. IBIS - Interoperability in Business Information Systems 2(2), 97–117 (2006)

16. Stollberg, M., Roman, D., Toma, I., Keller, U., Herzog, R., Zugmann, P., Fensel, D.: Semantic web fred - automated goal resolution on the semantic web. In: Proc. of the 38th Hawaii International Conference on System Sciences (2005)
17. Studer, R., Benjamins, R., Fensel, D.: Knowledge engineering: Principles and methods. Data and Knowledge Engineering 25(1-2), 161–197 (1998)
18. The world bank's information solutinos group. About e-Government. Web, available (2007), http://www.worldbank.com/egov
19. United Nations. UN e-government survey 2008: From e-government to connected governance (2008), http://unpan1.un.org/intradoc/groups/public/documents/UN/UNPAN028607.pdf
20. Valencia-García, R., Castellanos-Nieves, D., Fernández-Breis, J.T., Vivancos-Vicente, P.J.: A methodology for extracting ontological knowledge from spanish documents. In: Gelbukh, A. (ed.) CICLing 2006. LNCS, vol. 3878, pp. 71–80. Springer, Heidelberg (2006)
21. Wooldridge, M.: An introduction to MultiAgent Systems. Ed. John Wiley & Sons Ltd., Chichester (2002)
22. Zhao, L., Mehandjiev, N., Macaulay, L.: Agent roles and patterns for supporting dynamic behavior of web services applications. In: Proc. of the 3rd International Conference on Autonomous Agents and Multi-Agent Systems, New York, USA (2004)

Towards a Broker Agent
in the Semantic Services Environment

Özgür Gümüs, Önder Gürcan, and Oguz Dikenelli

Ege University, Department of Computer Engineering, Bornova, 35100 Izmir, Turkey
{ozgur.gumus,onder.gurcan,oguz.dikenelli}@ege.edu.tr

Abstract. Brokers provide important discovery and synchronization mechanisms among autonomous agents. Their mediation and coordination properties make brokers natural candidate components for the semantic services environment where each service is described, discovered and accessed semantically. However, brokers with rich functionality of discovery and mediation are not part of the current semantic services environment. This paper discusses the design considerations of a broker agent in the semantic services environment. In this context, the required services of this broker agent are analyzed, a multi-agent system infrastructure including such a broker agent for the semantic services environment is proposed and an interaction protocol, based on FIPA specifications, for brokerage in this environment is given.

1 Introduction

Semantic web services are described and accessed using ontologies. Usage of these ontologies requires a service environment which provides discovery and execution capabilities. Such environments have been extensively studied in the literature based on OWL-S[1] and WSMO[2]. It has been accepted that brokers are critical and useful architectural entities for such semantic service environments[13,2,11].

Brokers provide coordination and mediation mechanisms when there is a need to facilitate the interaction between two or more parties. For example, if two parties want to communicate, but they do not share a common language, brokers may provide translation services. Furthermore, they may provide anonymization for one or both of the parties, by mediating the transaction. Brokers are also one of the main discovery and synchronization mechanisms among autonomous agents [9,3,14]. Because of their mediation and coordination properties as well as wide applicability, brokers are a natural candidate component for the semantic services environment.

In this paper, design considerations of a broker agent in the semantic services environment which is based on the Semantic Web Services Initiative[3] Architecture (SWSA) abstract architecture are discussed. Main focus of our design is to

---

[1] Semantic Markup for Web Services, http://www.daml.org/services/owl-s/

[2] Web Service Modelling Ontology, http://www.wsmo.org/

[3] Semantic Web Services Initiative (SWSI), http://www.swsi.org/

R. Kowalczyk et al. (Eds.): SOCASE 2008, LNCS 5006, pp. 31–44, 2008.
© Springer-Verlag Berlin Heidelberg 2008

provide a flexible and reusable broker architecture conforms to this abstract architecture. SWSA defines three consecutive phases for service usage: discovery, engagement and enactment. Each phase includes different activities to satisfy internal requirements of that phase. The critical point here is that each activity can be implemented in a different way based on the application requirements. For example, service selection activity of the discovery phase can be executed in various ways such as reputation based service selection or semantic similarity based. In this context, our broker agent provides such a flexible architecture in which different implementations of activities can be plugged-in to the system easily. Also an enhanced interaction protocol between the broker agent and other agents based on FIPA specifications is defined within the paper.

The rest of the paper is structured as follows: in Section 2 a background presenting the foundation of our work is given. Section 3 represents an MAS infrastructure for semantic service brokerage. The broker agent is introduced in Section 4, and Section 5 gives the brokering protocol which is used in this infrastructure. Section 6 gives an illustrative example and finally Section 7 concludes the paper and discusses the future work.

2 Background

There are some standardization efforts for semantic web services that allow web services to be able to work in the semantic web environment. The most attractive ones are OWL-S and WSMO. OWL-S is an ontology system for describing web services. It consists of a profile ontology to advertise the capabilities of a service, a process ontology to describe the functionality and composition of a service and a grounding ontology to give details of how to access a service. OWL-S is the first effort for the semantic web services concept but it's not a complete system and meaning of its some elements is not clearly defined. On the other hand, WSMO meta-model describes four top level elements: ontologies, goals, web services and mediators. WSMO is said to be more complete framework but it's not based on W3C standards such as OWL and SWRL[4]. Also it does not make use of OWL ontologies and it looks like a workflow system in a distributed and heterogeneous service environment.

Meanwhile the Semantic Web Services Initiative Architecture Committee[5] has created a set of architectural and protocol abstractions that serve as a foundation for semantic web service technologies [1]. The proposed SWSA framework builds on the W3C Web Services Architecture working group recommendation[6] and attempts to address all requirements of semantic service agents: dynamic service discovery, service engagement, service process enactment and management, community support services, and quality of service (QoS). This architecture is based on Multi-Agent System (MAS) infrastructure because the specified requirements

---

[4] Semantic Web Rule Language, http://www.w3.org/Submission/SWRL/

[5] SWSI Architecture Committee, http://www.daml.org/services/swsa/

[6] W3C Web Services Architecture Working Group, Web Services Architecture Recommendation, 11 February 2004, http://www.w3.org/TR/ws-arch/

can be accomplished with asynchronous interactions based on predefined protocols and using goal oriented software agents.

The SWSA framework describes the overall process of discovering and interacting with a semantic web service in three consecutive phases. The first phase, called candidate service discovery, is searching for available services that can (potentially) accomplish some set of a client agent's internal goals or objectives. The second phase, called service engagement, includes the process of interpreting candidate web service enactment constraints and then negotiating with prospective services until reaching an agreement. The following phase is the service enactment which completes mutually agreed upon objectives of client and service by following the service's published protocols. The client provides required inputs for the service to be executed and knows what to do whether service succeeds or not. The SWSA framework also determines the actors of each phase, functional requirements of each phase and the required architectural elements to accomplish these requirements in terms of abstract protocols.

There are some broker applications for semantic service environments such as IRS-III framework[2] and a broker for OWL-S web services[11]. IRS-III describes a framework which is based on SESA architecture[13]. In SESA, the middleware layer called as Semantic Execution Environmet is the core of the architecture which defines the conceptual functionality that is imposed on the architecture. Each such functionality could be realized by a number of so called middleware services. IRS-III is a reference implementation of SESA that takes a semantic broker based approach to create applications from semantic web services by mediating between a service requester and one or more service providers. Paolucci et. al.[11] provides a broker architecture and an implementation based on OWL-S. This architecture defines required functionalities and dependent protocols to manage OWL-S based service usage process.

Unlike the aforementioned studies, we took SWSA as a base architecture because it covers all aspects (including negotiation, agreement, trust etc.) of semantic service environment in a conceptual level. When the functionalities of the semantic service enviroment are described in a comprehensive manner as in the SWSA, agent based approaches are the only way to implement such an environment as defined in SWSA. Our broker is developed as an agent that supports SWSA phases and has a pluggable infrastructure to implement activities of those phases.

3   An MAS Infrastructure for Semantic Service Brokerage

In order to make brokerage real in the semantic services environment, we introduce an MAS infrastructure (based on our previous works [6,7]) which fulfills fundamental requirements of the conceptual model of SWSA. To realize this infrastructure in a reasonable way, we have some assumptions explained below:

- There are platform ontologies that represent the application domain(s) of the platform. These ontologies are designed by platform's administrator and, stored and managed by platform's ontology agent.

- The possible services that are allowed to be provided within the platform are represented as goal templates. So service provider agents obey these templates for service provision and service requester agents use them to express their service needs. These goal templates are described similar to semantic service descriptions (inputs, outputs, preconditions and effects) by platform's administrator and stored in broker agent. A similar approach is used in [12].

All agents in this infrastructure have the following general capabilities:

- They can utilize semantic web technologies to represent and manipulate knowledge.
- They publish semantic descriptions of their service capabilities and execution details to enable prospective consumers can interpret when selecting and invoking these services.
- They represent their service descriptions as instances of predefined goal templates (goal instance) of the platform.
- They define their service requests by using goal templates. These requests must be relevant to platform's goal templates.
- There are predefined generic plans for each phase of SWSA. Each agent can use these plans using the planner introduced in [5,7].

The infrastructure presents an IEEE FIPA[7] compliant MAS which includes the following main components: Broker Agent, Service Provider Agent, Service Requester Agent and Ontology Agent. Communication between these agents takes place according to the well known Agent Communication Language (ACL)[8].

Broker Agent (BA) is the directory facilitator of the proposed MAS. It performs both discovery and mediation between requester agents and provider agents of semantically described services in the platform. In detail, BA stores the advertised service capabilities (goal instances) of the provider agents in its Service Repository. It can perform semantic service matching between a requested goal (defined as a goal template) and advertised services in order to determine semantically most appropriate services for a request. To do this, BA employs a Semantic Service Matcher that retrieves goal instances for a given goal template using a semantic similarity metric[10]. On the other hand, BA also stores predefined goal templates of the platform in a repository called Goal Template Repository (GTR). Services advertised by the platform agents must conform to these templates. Also these templates are used by agents in order to express their service needs.

Service Provider Agent (SPA) plays role of a service provider. SPAs are wrapper agents which realize inclusion of external web services into the platform depending on the SWSA phases. External web services could be either pure web services (WSDL) or semantic web services (OWL-S, WSMO etc.). While invoking the external services, SPA manages required conversions on service inputs

---

[7] Institution of Electrics and Electronics Engineers (IEEE) Foundation for Intelligent Physical Agents, http://www.fipa.org/

[8] FIPA Agent Communication Language Specifications, http://www.fipa.org/repository/aclspecs.html

and outputs in two directions using goal instance of requests. Details of SPA's internal architecture is out of scope of this paper.

Service Requester Agent (SRA) is a service client agent that needs services of other agents to achieve its objectives. When SRA needs a service, it first defines its request by creating a goal instance of a platform's goal template, the semantic similarity level and the required inputs for this template. Finally it delegates selection and execution of appropriate service(s) to BA.

Ontology Agent (OA) includes an Ontology Repository in which ontologies used in the platform and mappings between them are stored. OA provides contolled access and query on these ontologies and translation services for other members of the platform.

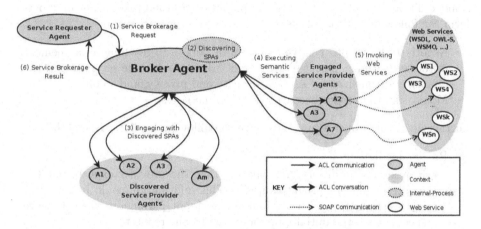

Fig. 1. Overall scenario of semantic service brokerage on the proposed MAS infrastructure

Figure 1 shows the overall scenario of semantic service brokerage on this MAS infrastructure. SRA starts the process by sending a service brokerage request to Broker Agent (step 1). Then Broker Agent searches for available SPAs that can (potentially) accomplish some set of a SRA's internal goals (step 2). After that, Broker Agent tries to make service engagement with these discovered agents (step 3). The following that the broker does the service enactment (steps 4, 5) which completes mutually agreed upon objectives of SRA and SPAs. Finally the broker collects the results and returns a response to SRA (step 6).

4   Broker Agent

4.1   Requirements

As described in Section 3, a broker agent in the semantic services environment is expected to perform both discovery and mediation between requester agents and provider agents of those services. So it must perform the following basic tasks:

- Registration of the capability advertisements of service providers.
- Interpretation of requesters' service requests (goals) that must be fulfilled by a service provider and saving the input parameters for later use while executing the engaged service.
- Discovering the candidate services and selecting the best one(s) based on the requester's service request.
- The engagement with the selected service provider(s).
- Invocation of the engaged service(s) on behalf of the requester and interacting with the service provider(s) as necessary.
- Returning the service execution results to the requester.

On the other hand, the accomplishment of the aforementioned tasks requires some additional tasks and mediation capabilities. So the broker agent may perform the following additional tasks:

- Negotiating with prospective service providers until reaching an agreement in the service engagement phase (Negotiations can include service price, the quality and timeliness of service, security and privacy, and so on).
- Monitoring the service execution for QoS metrics.

Furthermore, the broker agent must have the following mediation capabilities:

- Process mediation:
  - Provision of additional information that the provider expects but the requester did not provide initially. This information could be an extra input parameter or an attribute of an input parameter which is more specific than the initially provided one by the requester.
  - Elimination of extra information (input) that the requester is provided but the provider does not expect to perform the service.
  - Elimination of extra information (output) that is resulted from the execution of the service but the requester does not expect.
- Functional mediation: Dynamic service composition using advertised services when no single service meets the requester's need or goal.

Also brokers are expected to perform translation between the information produced and/or consumed by requesters and providers if they use different ontologies (data mediation). But, since this mediation is managed by the SPAs and all advertised services will conform to platform's goal templates in the proposed platform, the broker agent does not make data mediation.

4.2   Internal Architecture

The architecture of the developed broker agent is built on the top of SEAGENT[4]'s agent architecture and is composed of two layers: agency and broker (Figure 2). *Agency* layer includes the core agent modules. SEAGENT's core agent architecture is composed of three main modules: Messaging Service (MS), Goal Manager (GM) and Planner. MS manages the communication of

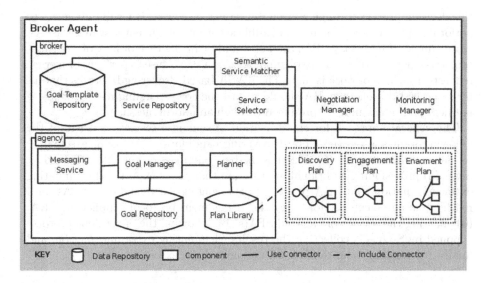

Fig. 2. Simplified Broker Agent Architecture

the agent, extract goals from requests and gives them to GM. GM manages the internal goals of the agents and is responsible for finding the best plan to achive a goal using the internal Goal Repository. Planner is at the hearth of the agent and controls the behaviours of the agent using plans in the Plan Library with respect to the goals.

Broker layer, on the other hand, includes brokerage related components. These components are brokering plans, generic modules (which are used by brokering plans), Goal Template Repository and Service Repository. The generic modules provide a pluggable architecture in which they can be changed according to application domain and the strategies used. Shortly these modules are Semantic Service Matcher (SSM), Service Selector (SS), Negotiation Manager (NM), Monitoring Manager (MM). Broker agent mainly has two goals which are named as *service brokerage* and *service registration*. *Service brokerage* goal includes three sub-goals for each phase of SWSA. These sub-goals are defined as seperate plans executed by the Planner (Figure 2). The details of these plans are given in [7]. *Discovery Plan* controls the discovery phase using SSM for capability matching and SS for service selection. *Engagement Plan* is used to control the engagement phase. It uses NM to cope with the difficulties of negotiation. *Enactment Plan* enacts the engaged services and monitors their process of execution using MM.

4.3 Brokering Process

When requesting the execution of a specified service through brokerage, SRAs interact with BA through the SWSA-based Brokering interaction protocol (explained in Section 5). According to this protocol, when BA receives a service execution request, it can either agree or refuse to perform the action. BA performs this decision process through a semantic service discovery. BA will only agree to perform a specific execution if it is possible according to currently available SPAs. For example, if necessary SPAs are not found or in the case of lack of input parameters, then BA refuses to perform it. Also, BA can refuse to perform brokerage requests if its load of work is high. The SWSA-based Brokering IP states that after a successful brokerage, BA should return the corresponding results or just a notification if no results are produced through an inform message.

The general approach for the execution of semantic services consists of the following sequence of steps: (1) discovering candidate SPAs using a semantic service matching engine, (2) selecting the best one(s) using the info provided by SRA, (3) provisioning additional information that SPAs expect, (4) eleminating extra information that SRA provided, (5) engaging with candidate SPAs until reaching an agreement by following the engagement protocol supplied by SRA, (6) enacting the service of engaged SPA(s) by following the enactment protocol supplied by SRA, (7) collecting the results, eleminating extra information, and finally, (8) sending them to SRA who requested the execution.

5   SWSA-Based Brokering Interaction Protocol

As discussed above, generally speaking, a broker is an agent that offers a set of communication facilitation services to other agents using some knowledge

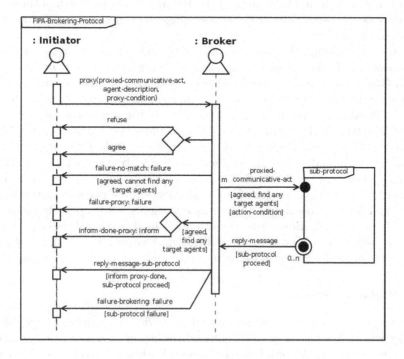

Fig. 3. FIPA Brokering Interaction Protocol

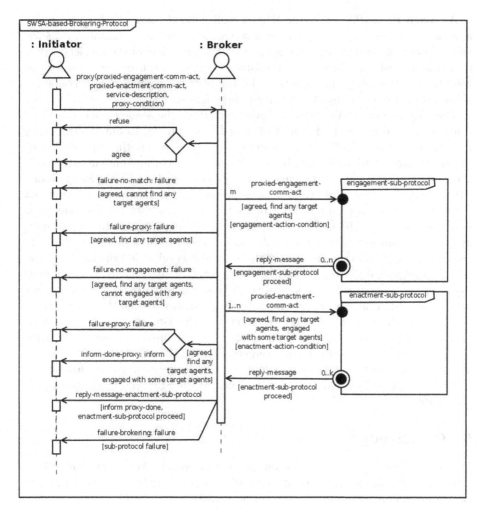

Fig. 4. SWSA-based Brokering Interaction Protocol for the semantic services environment

about the requirements and capabilities of those agents. FIPA has an interaction protocol (IP) to support brokerage interactions in mediated systems and multi-agent systems[9]. The representation of the FIPA Brokering Interaction Protocol is given in Figure 3. This protocol is a macro IP since the *proxy* communicative act[10] for brokerage embeds a communicative act as its argument and so the IP for the embedded communicative act is also embedded in this IP. This embedded IP (sub-protocol) guided some parts of the remainder of the interaction.

---

[9] FIPA Brokering Interaction Protocol Specification, www.fipa.org/specs/fipa00033/
[10] FIPA Communicative Act Library Specification, www.fipa.org/specs/fipa00037/

Although the FIPA Brokering IP is a well defined macro IP for brokerage, it is not quite suitable for agents in a semantic services environment. In such an environment, agents need to engage with the providers of the service(s) that they will request. In a brokerage, the Broker should do this engagement for the Initiator. So the proxy message that the Initiator sends, should also contain a sub-protocol for engagement and criteria for engagement.

Thus we defined a SWSA-based Brokering IP for the semantic services environment by extending the FIPA Brokering IP (Figure 4). In our IP, the *proxy message* contains the following: a *referential expression* denoting the services of the target agents to which the broker should forward the communicative act, the *communicative act for the engagement process* (such as FIPA Query-Reply, FIPA Request or negotiate-commitment protocols [1]), the *communicative act for enactment* to forward and a set of *proxy conditions* containing *engagement and enactment conditions* such as maximum number of agents to engage with and QoS parameters. Once the Broker has agreed to be a *proxy*, it discovers the candidate (target) agents using the semantic service description from the *proxy message* and the *proxy-condition* parameter. If such agents can be found, say k agents, the Broker selects m of them ($m <= k$) using the selection creteria provided by SRA, and begins m engagement interactions using the *engagement-sub-protocol*. At the end of these interactions, the Broker engages with zero or more target agents, say n. If there are engaged provider agents, the Broker begins one or more enactment interactions using the *enactment-sub-protocol*. When the *enactment-sub-protocol* completes, the Broker forwards the final *reply-message* (after making required eleminations) to the SRA and the brokering IP terminates.

6   Case Study

In order to demonstrate the precision for the proposed broker agent, this section discusses it for an example case in the tourism domain (which is a very popular case study domain in semantic web services area).

6.1   Scenario

Suppose we have a MAS platform for the tourism domain. Within this platform there are mainly three kinds of agents: traveller agents, tourism agents, and a broker agent. Traveller agents behaves on behalf of their users to help them for planning their travels. Tourism agents provides tourism related services to the platform agents such as finding best destinations for activities (surfing, skiing etc.), finding best accommodations to reside etc. Broker agent is responsible for the mediation between traveller agents and tourism agents within the platform.

For example, Özgür wants to surf in a specific place during his holiday. The places where people can surf are described according to the types of waves on the sea. Waves are classified according to their intensities such as shape and speed[11]. Wave tube shape is defined by its length to width ratio (Figure 5).

---

[11] http://en.wikipedia.org/wiki/Surfing

When width exceeds length, the tube is described as *square* (ratio is <1:1). When wave tube has a ratio of 1-2:1, the tube is described as *round.* Otherwise the tube is described as *almond* (ratio is >2:1). Wave tube speed is defined by angle of peel line. If peel line is 30° then the tube is described as *fast*, if it is 45° it is defined as *medium* and if it is 60° the tube is defined as *slow.* In Table 1, sample destinations for surfing according to wave intensities are shown.

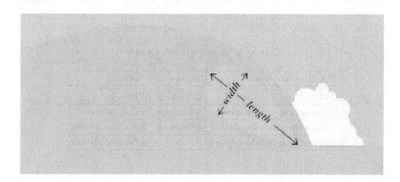

Fig. 5. The geometry of tube shape can be represented as a ratio between length and width

Özgür wants to travel a surfing destination where wave tube shape is *almond* and wave tube speed is *medium.* So, he directs his agent to find best available places. It is expected for this agent to find *Jeffreys Bay* and *Bells Beach* as destinations (see Table 1).

Table 1. Wave Intensity Table

	Fast	Medium	Slow
Square	The Cobra	Teahupoo	Shark Island
Round	Speedies, Gnaraloo	Banzai Pipeline	
Almond	Lagundri Bay, Superbank	Jeffreys Bay, Bells Beach	Angourie Point

6.2 Implementation

The Broker Agent was implemented by using the SEAGENT agent platform [4] which is a goal-oriented and semantic web enabled MAS framework. Within the developed Broker Agent, semantic services are described using OWL-S and thus, for capability matching, OWLS-MX tool [8] is used. For sample semantic services, the service retrieval test collection OWLS-TC2.2[12] (which is developed for OWLS-MX tool) is used. This test collection is preferred because it provides a large number of services from several domains, test queries and relevant ontologies (1007 OWL-S 1.1[13] services from seven domains -including travel

---

[12] http://projects.semwebcentral.org/frs/download.php/373/owls-tc2_2_rev_1.zip
[13] http://www.daml.org/services/owl-s/1.1/overview/

domain- and 29 test queries). We have chosen 4 of 107 service profiles in travel domain from OWLS-TC2.2 to resemble the services in the example. *surfing_destination* service description is used as the goal template of the SRA. Services provided by SPAs are *surfing_beach, surfing_destination, surfing_nationalpark* and *activity_nationalpark*. Table 2 shows providers, names, inputs and outputs of these provided services. And Table 3 shows the semantic similarity levels (match degree - explained in [8]) of them with respect to the *surfing_destination* goal template.

Table 2. Semantic Services

SPA	Service Name	Inputs	Outputs
spa1	surfing_beach	#Surfing	#Beach
spa2	surfing_destination	#Surfing	#Destination
spa3	surfing_nationalpark	#Surfing	#NationalPark
spa4	activity_nationalpark	#Activity	#NationalPark

The mechanism is as follows. Each SPA are registered to *broker_agent* (BA) with their services. Then a *traveller_agent* (SRA) that aims to know about the destination for windsurfing prepares a service brokerage request (proxy message in Figure 6, messages are shown in well known FIPA ACL String Representation[14]). This message contains the goal template of the requested service (*goal-template*), a semantic similarity degree (*match-degree*), protocols for engagement and enactment phases (*protocols*) and selection criteria for these phases (*selection-criteria*). *traveller_agent* send this message to *broker_agent* using the proposed SWSA-based Brokering IP.

Table 3. Semantic similarity levels

Service Name	Match Degree
surfing_beach	Plug-in
surfing_destination	Exact
surfing_nationalpark	Subsumes
activity_nationalpark	Fail

broker_agent discovers suitable services using *surfing_destination* goal template with the help of OWLS-MX tool. At the end of this discovery, *broker_agent* understands that *spa1, spa2* and *spa3* are suitable candidate SPAs. Then *broker_agent* starts an engagement process with these 3 SPAs and reaches agreements with *spa2* and *spa3*. After that *broker_agent* chooses the SPA (*spa2*) that provides the semantically most appropriate service and starts an enactment process with it. Finally it collects the results from *spa2* and forwards an inform

---

[14] FIPA ACL Message Representation in String Specification,
http://www.fipa.org/specs/fipa00070/

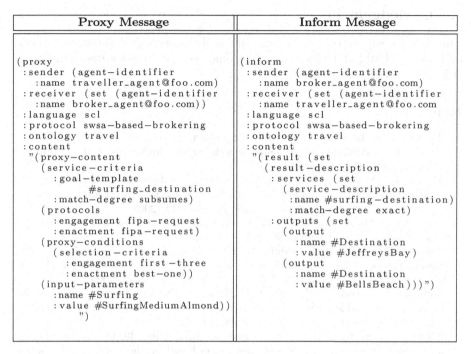

Proxy Message	Inform Message
``(proxy
:sender (agent-identifier
:name traveller_agent@foo.com)
:receiver (set (agent-identifier
:name broker_agent@foo.com))
:language scl
:protocol swsa-based-brokering
:ontology travel
:content
"(proxy-content
(service-criteria
:goal-template
#surfing_destination
:match-degree subsumes)
(protocols
:engagement fipa-request
:enactment fipa-request)
(proxy-conditions
(selection-criteria
:engagement first-three
:enactment best-one))
(input-parameters
:name #Surfing
:value #SurfingMediumAlmond))
")```	```(inform
:sender (agent-identifier
:name broker_agent@foo.com)
:receiver (set (agent-identifier
:name traveller_agent@foo.com
:language scl
:protocol swsa-based-brokering
:ontology travel
:content
"(result (set
(result-description
:services (set
(service-description
:name #surfing-destination)
:match-degree exact)
:outputs (set
(output
:name #Destination
:value #JeffreysBay)
(output
:name #Destination
:value #BellsBeach)))")```

Fig. 6. Proxy Message and Inform Message

message (inform message in Figure 6) that contains the details of the service executed (*:services*) and the results of this service (*:outputs*) -which are destinations for windsurfing (Jeffreys Bay and Bells Beach)- to *traveller_agent*.

7   Conclusion and Future Work

In this paper we have shown how design of a broker agent in the semantic services environment could be. In particular, we have introduced an MAS infrastructure involving such a broker agent. Then we put forward the requirements of this broker agent. Furthermore we have shown that FIPA brokering interaction protocol is not sufficient for a SWSA based semantic service environment and thus we have extended this interaction protocol for brokerage within the environment.

As a future work we first will enhance the evaluation of the broker agent using the introduced protocol within the proposed MAS infrastructure. This enhanced evaluation will be based on the case study given in this paper. After this evaluation, we will study to extend the brokering interaction protocol in respect to provision of additional input data that the provider expects and monitoring of the enactment sub-protocol as a next step. And also dynamic service composition using the advertised services when no single service meets the requester's needs will be studied.

Acknowledgements

This work is supported by the Scientific and Technological Research Council of Turkey (TÜBİTAK) Electrical, Electronics and Informatics Research Group (EEEAG) under grant 106E008.

References

1. Burstein, M., Bussler, C., Zaremba, M., Finin, T., Huhns, M.N., Paolucci, M., Sheth, A.P., Williams, S.: A semantic web services architecture. IEEE Internet Computing 9(5), 72–81 (2005)
2. Cabral, L., Domingue, J., Galizia, S., Gugliotta, A., Tanasescu, V., Pedrinaci, C., Norton, B.: IRS-III: A broker for semantic web services based applications. In: International Semantic Web Conference, pp. 201–214 (2006)
3. Decker, K., Williamson, M., Sycara, K.: Matchmaking and brokering. In: Proceedings of the Second International Conference on Multi-Agent Systems (1996)
4. Dikenelli, O., Erdur, R.C., Gümüs, Ö., Ekinci, E.E., Gürcan, Ö., Kardas, G., Seylan, I., Tiryaki, A.M.: Seagent: A platform for developing semantic web based multi agent systems. In: AAMAS, pp. 1271–1272. ACM, New York (2005)
5. Ekinci, E.E., Tiryaki, A.M., Gürcan, Ö., Dikenelli, O.: A planner infrastructure for semantic web enabled agents. In: Meersman, R., Tari, Z. (eds.) OTM 2007. LNCS, vol. 4804, pp. 95–104. Springer, Heidelberg (2007)
6. Gümüs, Ö., Gürcan, Ö., Kardas, G., Ekinci, E.E., Dikenelli, O.: Engineering an mas platform for semantic service integration based on the swsa. In: Meersman, R., Tari, Z. (eds.) OTM 2007. LNCS, vol. 4804, pp. 85–94. Springer, Heidelberg (2007)
7. Gürcan, Ö., Kardas, G., Gümüs, Ö., Ekinci, E.E., Dikenelli, O.: An MAS Infrastructure for Implementing SWSA based Semantic Services. In: Huang, J., Kowalczyk, R., Maamar, Z., Martin, D., Müller, I., Stoutenburg, S., Sycara, K.P. (eds.) SOCASE 2007. LNCS, vol. 4504, pp. 118–131. Springer, Heidelberg (2007)
8. Klusch, M., Fries, B., Sycara, K.: Automated semantic web service discovery with owls-mx. In: AAMAS 2006: Proc. of the 5th Inter. joint Conf. on Autonomous agents and multiagent systems, NY, USA, pp. 915–922. ACM Press, New York (2006)
9. Klusch, M., Sycara, K.: Brokering and matchmaking for coordination of agent societies: a survey, pp. 197–224 (2001)
10. Paolucci, M., Kawmura, T., Payne, T., Sycara, K.: Semantic matching of web services capabilities. In: First Int. Semantic Web Conf. (2002)
11. Paolucci, M., Soudry, J., Srinivasan, N., Sycara, K.P.: A broker for owl-s web services. Extending Web Services Technologies: The Use of Multi-Agent Approaches, 79–98 (2004)
12. Stollberg, M., Roman, D., Toma, I., Keller, U., Herzog, R., Zugmann, P., Fensel, D.: Semantic web fred - automated goal resolution on the semantic web. In: 38th Annual Hawaii International Conference, p. 111 (2005)
13. Vitvar, T., Mocan, A., Kerrigan, M., Zaremba, M., Zaremba, M., Moran, M., Cimpian, E., Haselwanter, T., Fensel, D.: Semantically-enabled service oriented architecture: Concepts, technology and application. In: Service Oriented Computing and Applications. Springer, London (2007)
14. Wong, H., Sycara, K.: A taxonomy of middle-agents for the internet. In: ICMAS 2000 (2000)

Pattern-Based Semantic Tagging
for Ontology Population

Masumi Inaba, Takayuki Iida, Tomohiro Yamasaki, Kosei Fume,
Yumiko Mizoguchi, Shinichi Nagano, and Takahiro Kawamura

Corporate Research & Development Center, Toshiba Corporation
1 Komukai-Toshiba-cho, Saiwai-ku, Kawasaki-shi, Kanagawa, 212-8582 Japan
masumi.inaba@toshiba.co.jp

Abstract. Ontology population has emerged as an increasingly important problem in semantic web services. In this paper, we propose a method using named entity recognition that extracts keywords from Web pages in order to populate a product ontology. The semantic classification determines meanings of terms and phrases by heuristic rules after the morphological analysis. In addition, our method classifies vocabularies into different semantic tags. Firstly, it records several lists of semantic tags to a history database. Then, we define some rules from the lists to extract a product name. Finally, the rules build and refine the product ontology semi-automatically. According to an evaluation, proposed method achieved 87.1% precision and 87.4% recall. Thus, it can suggest some instances, and it decreases cost of updating the ontology.

1 Introduction

In Service-Oriented Computing, semantics-based applications are being actively developed [1-6]. Elgedawy et al. introduced several new concepts as a semantic-based service composition for web services [1]. Also, Ubiquitous Service Finder (USF), in which user can invoke the services semantically [2]. In these cases, the ontology is used as a matching technique for the semantic web services. The ontology is a basic knowledge and would be important in Service-Oriented Computing to bind the user's situation and the services. However, each application needs to construct an initial ontology. Further, the application continues to update the ontology. So, knowledge and skill concerning the ontology are required.

On the other hand, almost everyone is able to create Web content easily by means of Content Management Systems (CMS) including Weblog, Wiki, and Social Networking Services (SNS). Consumers actively engage in "word of mouth" communication about products and services. This is called Consumer-Generated Media (CGM). CMS generates formalized data, RSS and HTML. We have developed a reputation extraction system from Weblogs, Ubiquitous Metadata Scouter (UMS) [6]. It gathers lots of Weblogs from the Web and analyzes the reputation of a product by referring to a product ontology. The product ontology is constructed of classes, instances, and relations between them: is-a, instance-of, etc. There are lots of product descriptions, which keep increasing every day. Therefore, it is necessary to collect the product description from the web periodically. However, it is difficult to construct the

R. Kowalczyk et al. (Eds.): SOCASE 2008, LNCS 5006, pp. 45–55, 2008.

product ontology manually, because the collected product information includes lots of noisy vocabularies. For example, a spec description and an explanation of a product are noisy vocabularies. These vocabularies include a product metadata, however, can not use such as a product name. In this way, the noisy vocabularies such as the spec description and the explanation of the product are deleted from the product description, and then we can get a product name.

Semantic tagging is a text analysis technique, which puts a label to a vocabulary in a text. Firstly, the vocabularies are parsed by the morphological analysis. Next, a part of speech is categorized with the semantic tags, which are defined in advance. For example, a person name includes an athlete and a singer. Also a location includes a country name and a region name. As a corpus to gather the product names, we use large product information by Electric Commerce Site API. It is categorized with the product type and good corpus than general text includes lots of noisy description. Firstly, we record pattern knowledge from this corpus to a database. Next, we consider get a paraphrase and an abbreviation.

We propose a way of generating a product ontology automatically from Web pages. Named entity recognition combines morphological analysis with semantic classification [7, 8], extracts a name of a person, a place, date, etc. The semantic classification extracts the meanings of terms and phrases by the heuristic rules. In addition, our method classifies vocabularies into different semantic tags as they are collected. It records several lists of semantic tags as a history to the database. Firstly, our method parses large product information, and records a pattern of the semantic tagging. The pattern includes a list of the semantic tags, the number of the semantic tags which constructed an instance, and a frequency score of the semantic tags. We name the list of the semantic tags "array". Then, referring to the database, it classifies new vocabularies. The pattern applies to similar product information including new description. The product information is a corpus which includes lots of candidates for an instance in a concept. The semantic tagging uses the pattern of the array, and puts some semantic tags to the vocabularies in the corpus. Then it determinates availability for an instance. Available instance is recorded to the database as a new pattern.

Thus, our proposed Named entity recognition method acquires instances efficiently in a lightweight ontology. The aim of automatic construction technology for the ontology is to decrease generation cost of the large-scale ontology.

This paper is organized as follows. Section 2 describes related works. Next, Section 3 considers our Named entity recognition method. In section 4, an evaluation is reported based on an experiment on the proposed technique. Finally, in section 5, a summary is presented and future problems are identified.

2 Ubiquitous Metadata Scouter

Our reputation extraction system, Ubiquitous Metadata Scouter, is a semantic-based information retrieval of a product reputation from Weblogs. Firstly, it retrieves blog entries commenting on a specified product. Then, it extracts reputation expression and similar products from the retrieved blog entries. Next, it summarizes the reputation information on the target product. The main feature of the system is the semantic

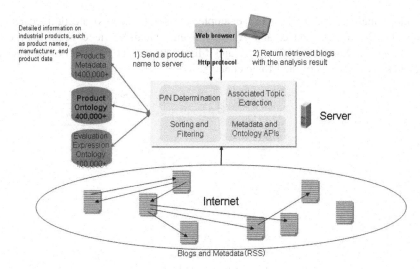

Fig. 1. Architecture of reputation extraction system

analysis of sentences in the retrieved blog entries using ontology, which is the description of concepts and their relationships. The ontology enables the following three features: P/N (positive/negative) determination of the product reputation, associated product extraction, and blog sorting and filtering by reputation relevance. Figure 1 shows the architecture of UMS. P/N determination is one of the text summarization techniques. It involves retrieving triples <subject, property, value> such as <car, speed, fast> for a target subject by checking modification relation through morphological analysis and syntactic parsing. Associated product extraction finds other products that are similar to a certain product. Finally, blog sorting and filtering extracts only useful articles. For example, articles to have lots of track backs, obvious positive opinions or strong negative opinions are selected. Then, the sales blog assumed that user's concern is relatively low in general are removed.

We have developed three databases, product metadata, product ontology, and evaluation expression ontology. The product metadata contains product descriptions; product names, manufacturers, specifications, prices. It contains more than one million products, each of which is described in a form of RDF. Second, product ontology contains more than 400 thousand products including DVDs, books, and electronics appliances. It is composed of an is-a relation representing mappings of individual products with product categories. Third, evaluation expression ontology contains sensitive expressions necessary for reputation analysis of blog sentences. Both ontologies are described in a form of OWL.

3 Pattern-Based Semantic Tagging

3.1 Ontology

Figure 2 described an ontology that we have constructed for a product. The product ontology has a class hierarchy classified by product genres and huge instances shown with individual product names. For example, classes (concepts) are defined as the genres, DVD, Movie, Music, and Animation etc. Then Movie concept has some product names, Movie A, Movie B, and Movie X etc. These product names are defined as the instances of the Movie concept.

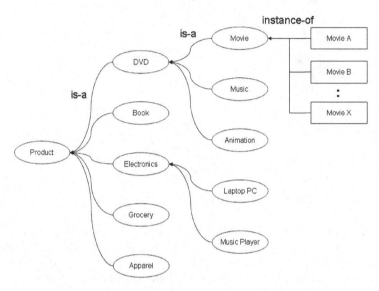

Fig. 2. Product ontology

Ontology population improves an instance of a target concept. It extracts a vocabulary from a corpus, and categorizes the vocabulary as an instance of the target concept.

3.2 Proposed Method

Figure 3 shows the proposed method which is composed of preprocessing, Morphological analysis, semantic tagging, and instance generator modules. A corpus consisted of aggregate of character string of a target concept. The corpus includes lots of candidates for getting an instance. Firstly, the preprocessing inputs the corpus which has relevant vocabularies of the target concept. Then, it normalizes the corpus. Secondly, Morphological analysis parses the corpus and extracts a part of speech, a verb, a noun, and an adjective etc. Thirdly, the semantic tagging refers a database for array of semantic tags, and gives the semantic tags to the vocabularies. Finally, the instance generator outputs a product name as an instance of the target concept by the heuristic rules. The rules are defined by a property of the semantic tag.

3.3 Examples

We illustrate some examples of our semantic tagging for ontology generation. Figure 1 described an ontology that we have constructed for a product. The product ontology

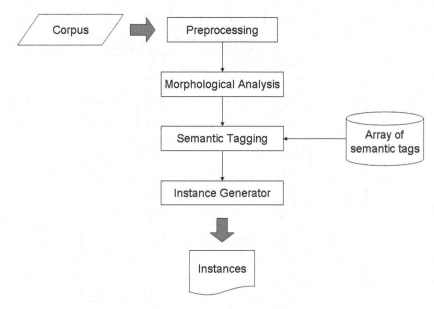

Fig. 3. Proposed Method

has a class hierarchy classified by product genres and huge instances shown with individual product names. The product name is expressed a proper noun on some rules [24, 25]. The product name can be used as a keyword for Web retrieval.

It is necessary to resolve the following two points in order to extract the product information from a Web page.

Firstly, we need to decide a target page. If the Web page is not decided, the Web retrieval is often used. However, a keyword is necessary to look for a pertinent page by the Web retrieval. Then, pertinent part is detected on the Web page. The keyword is used as a clue to look for the vocabulary that should be extracted from a huge text. The selection of an appropriate keyword is one of problems in the vocabulary extraction. We propose a new method for using a proper noun as a keyword for Web retrieval.

New product names are extracted from Web pages and the product ontology is constructed. Our proposed method uses arrays of semantic tags that is used to extract the proper noun from the product name.

A product has an identifiable name. For example, a name of apparel is constructed of several words: <Product, Size, Color>, <Manufacturer, Size, Product>, and <Manufacturer, Color, Product, Size>, etc. Similarly, a name of grocery is shown <Product, Weight, Pack>, <Manufacturer, Product, Flavor, Weight, Pack>, and so on. Then, the arrays make lists that are classified into the product genres.

An example of the arrays of semantic tags is shown in Table 1. In this case, the history data of the arrays contains three classes: A, B, and C. Then, an array is

Table 1. Array of semantic tags

Length	Array of semantic tags	Score (%)
1	<A>	50.0
	<B>	30.0
	<C>	20.0
2	<AB>	35.0
	<AC>	20.0
	<BA>	15.0
	<AA>	10.0
	<BC>	8.0
	<BB>	6.0
	<CA>	5.0
	<CB>	0.8
	<CC>	0.2
3	<ABC>	70.0
	<ABB>	20.0
	<BAC>	8.0
	<CAC>	2.0

composed of some semantic tags. In each array, it has the score that is calculated by the frequency of the semantic array. For instance, <ABC> is scored 70.0%. When an input word is the same array of three in length as the existing array of semantic tags, the score of the array of three in length is calculated and updated.

Pattern-based Semantic Tagging is shown in Figure 4. "Toshiba dynabook SS RX1" is separated into four vocabularies, "Toshiba", "dynabook", "SS", and "RX1". Then, these vocabularies are given semantic tags. The extracted vocabulary means "Manufacuturer", "Brand", "Series", and "Model". When the semantic tags are completed, a new array of semantic tags A of <Manufacturer, Brand, Series, Model> can be taken as an array of the Laptop PC. The new word is acquired from the array of semantic tags by repeating this process. For instance, "Toshiba dynabook Qosmio F40" is analyzed into "Toshiba", "dynabook", "Qosmio", and "F40". "Toshiba" and "dynabook" become existing data. On the other hand, "Qosmio" and "F40" are determined to unknown words. However, the array of semantic tags is matched. The input word is the same array of four in length as the existing array A. Referring to the history data of the array of semantic tags, <Manufacturer, Brand, *, *> matches a rule, <Manufacturer, Brand, Series, Model>. Therefore, "Qosmio" is extracted as a new word of "Series" and "F40" is known as a word of "Model".

Even if it is the same product classification, the array with a different length of the array and rules might be obtained. "Toshiba dynabook F40" are already known, it can be determined new array B that differs its length as shown in Figure 5. By repeating this work, information already-known increases. Thus, the accuracy of automatic processing improves.

Laptop PC

TOSHIBA dynabook SS RX1
TOSHIBA dynabook Qosmio F40
. . .

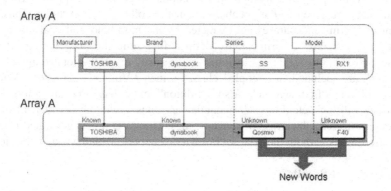

Fig. 4. Pattern-based Semantic Tagging

Laptop PC

TOSHIBA dynabook SS RX1
TOSHIBA dynabook Qosmio F40
TOSHIBA dynabook F40

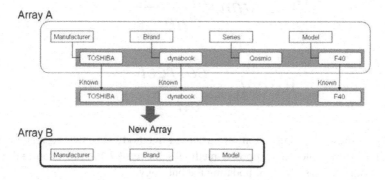

Fig. 5. New array extraction

4 Experiment

4.1 Experiment Overview

We evaluated the proposed method by using a DVD ontology in past seven years, including movies, music, cartoons, dramas and so on. We made correct answer data manually in 2 stages.

Firstly, DVD descriptions are collected from Electronic Commerce sites by hand. It is not an automatic collection. Data includes DVD data only. Secondly, the DVD title is made from the collected DVD descriptions. It corresponds a keyword for the Web retrieval.

We apply the proposed method to the collected DVD descriptions. Then, the semantic tags are given to the vocabularies in the collected DVD descriptions. The meanings of terms and phrases are extracted by the morphological analysis and the heuristic rules. In this experiment, we defined four semantic tags: Format, People, Volume, and Edition. "DVD" is included in "Format" class. "Actors/Actresses" and "Directors" are included in "People" class. Further, "Volume" class has "Vol. 1", "Season 1" and "first season". Also, "Edition" class has "special edition" and "Collector's Edition", etc. For each semantic tag, it has a rule to be extracted from the DVD descriptions by the arrays of semantic tags. The DVD title is generated from the DVD descriptions without these tags: "Format", "People", "Volume", and "Edition".

4.2 Experiment Results

The evaluation measure is how closer computed scores are to the correct answers. The proposed method was evaluated based on two criteria: precision and recall. Precision (1) denotes the ratio of correctly extracted terms over all extracted terms (all method). Recall (2) is defined as the number of correct answer divided by the total number of extracted terms from the corpus (all manual).

$$\Pr ecision = \frac{correct}{all_{method}} \tag{1}$$

$$\mathrm{Re}call = \frac{correct}{all_{manual}} \tag{2}$$

Table 2 shows a comparison of proposed method with the correct answers. According to the table, the proposed method achieved 87.1% precision and 87.4% recall. Thus, it decreases cost of updating the ontology.

Table 2. Precision and Recall

	Precision (%)	Recall (%)
Proposed Method	87.1	87.4

Table 3 shows matching rate of four rules. Edition rule is a highest rate in these rules for the DVD descriptions. Edition rule and Volume rule share over 80%. Therefore, we expect that Edition rule and Volume rule are effective to reduce amount of unnecessary word from the DVD title.

Table 3. Score of matching rules

Rule	Rate (%)
Edition	51.8
Volume	30.9
People	10.1
Format	7.2

5 Related Works

There are three techniques of the ontology generation:

1. Support environments such as manual construction
Protégé [9] is a popular ontology editor. It implements a rich set of knowledge-modeling structures and actions that support creation, visualization, and manipulation of various formats. OntoGen [10] is a semi-automatic ontology editor focusing on editing of a topic ontology. It integrates a text-mining technique and a machine learning into an efficient user interface lowering threshold for users who are not professional ontology engineers. During the ontology construction process, OntoGen suggests concepts and relations between the concepts. However, it does not suggest instances.

2. Semantic integration
This method shares data across disparate sources such as WordNet [11], Cyc [12], and EDR [13]. WordNet is a semantic vocabulary for the English language. It records the various semantic relations between these vocabularies, and Cyc is an ontology database of common sense knowledge. EDR is an electronic dictionary that catalogues the lexical knowledge of Japanese and English. It has unified thesaurus-like concept classifications with corpus databases. The concept classification dictionary is a sub-dictionary of the concept dictionary, which describes the similarity relation among concepts listed in the word dictionary. The semantic integration solves matching ontologies or schemas, detecting duplicate triples, reconciling inconsistent data values, modeling complex relations between concepts in different sources, and reasoning with semantic mappings [14].

3. Extraction from Web pages
This method constructs an ontology from web pages [15-17]. Tijerino et al. [16] use HTML tables. Further, Cohen et al. [17] propose a flexible learning system that uses not only tables but also lists in HTML. However, these methods require wrapper for each new pages.

Automation is indispensable for the construction of the large-scale ontology. The vocabulary that can be expanded depends on the amount and the quality of the resources. This is because a lot of new words are included in a new resource. Therefore, technique 3 is suitable for extracting new words.

There are two ways of vocabulary extraction from Web pages:

1. Vocabulary extraction from plain text
The vocabularies are extracted from a plain text. Then, some tags are deleted from the Web pages and morphological analysis is applied. Hearst [18] and Cimiano et al. [19] propose a vocabulary extraction combined with Web retrieval. Cimiano et al. present a method, C-PANKOW (Context-driven Pattern-based ANnotation through Knowledge On the Web), which has an advantage of no training required. Pasca et al. [20] propose a method in which two or more keywords are combined and the degree of similarity of the vocabulary determined.

2. Vocabulary extraction from structured data
This method extracts the meaning of the vocabulary from the document and the structure. If a target page is decided, we can use STALKER web wrapper [21]. In case of the target Web pages is not decided, its HTML document structure is analyzed by means of bootstrapping and the vocabulary is extracted. This method is proposed by Brin [22] and Agichtein et al. [23]. Brin proposes a method which called DIPRE. It exploits a pair between sets of patterns and relations to grow the target relation starting from a small database. For example, it extracts a relation of the pairs like <author, title> from HTML document. In other hand, Agichtein et al. propose Snowball system which extracts relations from large collections. However, these methods require training for each new scenario.

6 Conclusion

We have proposed a new method that generates a product ontology. It is attractive for acquiring a huge amount of fresh data by using Web resources. However, it is difficult to measure the accuracy of the acquired vocabulary automatically. The extraction accuracy can be improved by studying the measurement. For the aim of realizing the automatic ontology population, we introduced an approach to the vocabulary acquisition in which existing ontology is used. In our future research on the automatic ontology population, we intend to employ the Named entity recognition method.

References

1. Elgedawy, I., Tari, Z., Winikoff, M.: Exact functional context matching for web services. In: Proceedings of the 2nd international conference on Service oriented computing (ICSOC 2004) (2004)
2. Kawamura, T., Ueno, K., Nagano, S., Hasegawa, T., Ohsuga, A.: Ubiquitous Service Finder - Discovery of Services semantically derived from metadata in Ubiquitous Computing. In: Gil, Y., Motta, E., Benjamins, V.R., Musen, M.A. (eds.) ISWC 2005. LNCS, vol. 3729. Springer, Heidelberg (2005)
3. Sasajima, M., Kitamura, Y., Naganuma, T., Kurakake, S., Mizoguchi, R.: Task Ontology-Based Framework for Modeling Users' Activities for Mobile Service Navigation. In: Sure, Y., Domingue, J. (eds.) ESWC 2006. LNCS, vol. 4011, pp. 71–72. Springer, Heidelberg (2006)

4. Mizoguchi-Shimogori, Y., Nakamoto, T., Asakawa, K., Nagano, S., Inaba, M., Kawamura, T.: TV Navigation Agent for Measuring Semantic Similarity between Documents. In: Proceedings of 3rd International Workshop on Agents and Web Services in Distributed Environments (AWeSOMe 2007) (2007)
5. Cho, K., Kawamura, T.: BlogAlpha: Home Automation Robot using Ontology in Home Environment. In: Proceedings of Artificial Intelligence and Applications (AIA 2007) (2007)
6. Kawamura, T., Nagano, S., Inaba, M., Mizoguchi, Y.: Mobile Service for Reputation Extraction from Weblogs - Public Experiment and Evaluation. In: Proceedings of Twenty-Second Conference on Artificial Intelligence (AAAI 2007) (2007)
7. Punuru, J., Chen, J.: Learning for Semantic Classification of Conceptual Terms. In: IEEE International Conference on GRC 2007 (2007)
8. Liu, F., Zhao, J., Lv, B., Xu, B., Yu, H.: Product Named Entity Recognition Based on Hierarchical Hidden Markov Model. In: Proceedings of the Fourth SIGHAN Workshop on Chinese Language Processing (2005)
9. Protégé, http://protege.stanford.edu/
10. OntoGen, http://ontogen.ijs.si/
11. WordNet, http://wordnet.princeton.edu/
12. Cyc, http://www.cycfoundation.org/
13. EDR Electronic Dictionary, http://www2.nict.go.jp/r/r312/EDR/
14. Noy, N.F., Doan, A., Halevy, A.Y.: Semantic Integration. AI Magazine 26, 7–10 (2005)
15. Wong, T., Lam, W., Chen, E.: Automatic Domain Ontology Generation from Web Sites. Journal of Integrated Design & Process Science archive 9(3), 29–38 (2005)
16. Tijerino, Y.A., Embley, D.W., Lonsdale, D.W., Nagy, G.: Ontology generation from tables. In: Proceedings of the Fourth International Conference on Web Information Systems Engineering, pp. 242–249 (2003)
17. Cohen, W.W., Hurst, M., Jensen, L.S.: A flexible learning system for wrapping tables and lists in HTML documents. In: Proceedings of the 11th international conference on World Wide Web, pp. 32–241 (2002)
18. Hearst, M.A.: Automatic acquisition of hyponyms from large text corpora. In: Proceedings of the Fourteenth International Conference on Computational Linguistics, Nantes, France, pp. 539–545 (July 1992)
19. Cimiano, P., Ladwig, G., Staab, S.: Gimme' the context: context-driven automatic semantic annotation with C-PANKOW. In: Proceedings of the 14th international conference on World Wide Web May 10-14 (2005)
20. Pasca, M., Lin, D., Bigham, J., Lifchits, A., Jain, A.: Organizing and searching the world wide web of facts - step one: The one-million fact extraction challenge. In: Proceedings of the 21st National Conference on Artificial Intelligence (2006)
21. Muslea, I., Minton, S., Knoblock, C.A.: Hierarchical wrapper induction for semistructured information sources. Autonomous Agents and Multi-Agent Systems 4(1/2), 93–114 (2001)
22. Brin, S.: Extracting patterns and relations from the world wide web. In: Schek, H.-J., Saltor, F., Ramos, I., Alonso, G. (eds.) EDBT 1998. LNCS, vol. 1377, Springer, Heidelberg (1998)
23. Agichtein, E., Gravano, L.: Snowball: Extracting relations from large plaintext collections. In: Proceedings of the 5th ACM International Conference on Digital Libraries (2000)
24. Sakai, T., Saito, Y., Ichimura, Y., Koyama, M., Kokubu, T., Manabe, T.: ASKMi: A Japanese Question Answering System based on Semantic Role Analysis. In: RIAO 2004 Proceedings, pp. 215–231 (2004)
25. Frantzi, K., Ananiadou, S.: Extracting Nested Collocations. In: COLING 1996, pp. 41–46 (1996)

Service-Based Integration of Grid and Multi-Agent Systems Models

Clement Jonquet[1], Pascal Dugenie[2], and Stefano A. Cerri[2]

[1] Stanford Center for Biomedical Informatics Research (BMIR)
Stanford University School of Medicine
Medical School Office Building, Room X-215
251 Campus Drive, Stanford, CA 94305-5479 USA
jonquet@stanford.edu

[2] Laboratory of Informatics, Robotics, and Microelectronics of Montpellier (LIRMM)
National Center of Scientific Research (CNRS) & University Montpellier 2
161 Rue Ada, 34392 Montpellier, France
{dugenie,cerri}@lirmm.fr

Abstract. This position paper addresses the question of integrating GRID and MAS (Multi-Agent Systems) models by means of a service oriented approach. Service Oriented Computing (SOC) tries to address many challenges in the world of computing with services. The concept of service is clearly at the intersection of GRID and MAS and their integration allows to address one of these key challenges: the implementation of dynamically generated services based on conversations. In our approach, services are exchanged (i.e., provided and used) by *agents* through *GRID* mechanisms and infrastructure. Integration goes beyond the simple interoperation of applications and standards, it has to be intrinsic to the underpinning model. We introduce here an (quite unique) integration model for GRID and MAS. This model is formalized and represented by a graphical description language called Agent-Grid Integration Language (AGIL). This integration is based on two main ideas: (i) the representation of agent capabilities as Grid services in service containers; (ii) the assimilation of the service instantiation mechanism (from GRID) with the creation of a new conversation context (from MAS). The integrated model may be seen as a formalization of agent interaction for service exchange.

1 Introduction

The GRID and MAS communities believe in the potential of GRID and MAS to enhance each other because these models have developed significant complementarities [1]. One of the crucial explorations concerns the substitution by an agent-oriented kernel of the current object-oriented kernel of services available in Service Oriented Architectures (SOAs), including GRID. The Service Oriented Computing (SOC) community agrees that such a change will really leverage SOC scenarios by providing new types of services [2]. This key concept of *service* is clearly at the intersection of the GRID and MAS domains and thus may motivate an integration.[1] GRID is said to be the first

---

[1] [1] foresees services as the core 'unifying concept' that underlies GRID and MAS (also historically suggested by [3] and [4]).

R. Kowalczyk et al. (Eds.): SOCASE 2008, LNCS 5006, pp. 56–68, 2008.

distributed architecture (and infrastructure) really developed in a service-oriented perspective: Grid services are compliant Web services, based on the dynamic allocation of virtualized resources to an instantiated service [5]. Actually, GRID acquired major importance in SOA by augmenting the basic notion of Web Service with two significant features: service state and service lifetime management. Whereas Web services have instances that are stateless and persistent, Grid service instances can be either stateful or stateless, and can be either transient or persistent.[2] On the other hand, agents are said to be autonomous, intelligent and interactive entities who may use and provide services (in the sense of particular problem-solving capabilities). Actually, agents have many interesting characteristics for service exchange: they are reactive, efficient, adaptive, they know about themselves, they have a memory and a persistent state, they are able to have conversation, work collaboratively, negotiate, learn and reason to evolve, deal with semantics associated to concepts by processing ontologies, etc. MAS and SOC communities recently turned to one another considering the important abilities of agents for providing and using dynamic composed services, semantic services, business processes, etc. (see [7] for a recent overview of SOC challenges). Web services are often criticized because they are no more than Remote Procedure Calls (RPC) which have no user adaptation, no memory, no lifetime management, no conversation handling capabilities (simple request/answer interaction). They are passive, they lack semantics and they do not take into account the autonomy of components. The SOC community has realized that the notion of service has to surpass HyperText Transfer Protocol, current SOA standards (Web Service Definition Language (WSDL), Simple Object Access Protocol (SOAP), Universal Description Discovery and Integration (UDDI)), RPCs and eXtensible Markup Language (XML) to be enriched by results from other research domains such as information systems, concurrent systems, knowledge engineering, interaction and, especially, GRID and MAS.

To provide a service means to identify and offer a solution (among many possible ones) to the problem of another. The next generation of services will consist of dynamically generated services, i.e., services constructed on the fly by the service provider according to the conversation it has with the service user. In *Dynamic Service Generation* (DSG), term suggested by [8,9], the user (human or artificial) is not assumed to know exactly what the provider (also human or artificial) can offer him. He finds out and constructs step by step what he wants based on the service provider's reactions. The central idea of DSG is that a service may be based on a conversation. Actually, DSG highlights the idea of processing something new instead of merely delivering something that already exists. In everyday life, when somebody needs new clothes, *buying ready-to-wear clothes* is analogous to asking for a product, whereas *having clothes made by a tailor* is analogous to requiring a service to be generated. Singh and Huhns [7] talk about *service engagement*, instead of simple method invocation. In [8,9] we present the STROBE model as an agent representation and communication model designed and constructed in order to develop dynamically generated services. The shift from the currently limited perspective in service exchange scenarios to DSG is the topic addressed by this paper. It introduces a service based GRID-MAS integrated model to help to

---

[2] Grid service specifications are described both by Open Grid Service Architecture (OGSA) [5] and Web Service Resource Framework (WSRF) [6].

go towards this DSG vision by providing a common integration to help the community designing service architectures that benefits from both MAS and GRID interesting service features. In order to summarize our thoughts at the intersection of the three domains (GRID, MAS and SOC), we identify two key ideas:

- GRID and MAS have each developed a service oriented behaviour, therefore the concept of service may represent a common integration;
- New needs in service exchange scenarios are clearly highlighted and may be met by integrating GRID and MAS complementarities.

In [9,10], we introduce the *Agent-Grid Integration Language* (AGIL) as a GRID-MAS integrated systems description language which rigorously formalizes both key GRID and MAS concepts, their relations and the rules of their integration with graphical representations and a set-theory formalization. AGIL is both an integration model and a description language (i.e., a sort of UML for GRID-MAS integrated systems). In this position paper we present quickly the main ideas and principles of AGIL integration model.

2   Brief State of the Art

2.1   Brief GRID Overview

The GRID aims to enable *flexible, secure, coordinated resource sharing and coordinated problem solving in dynamic, multi-institutional virtual organization.* Actually, it was originally designed to be an environment with a large number of networked computer systems where computing (Grid computing) and storage (data Grid) resources could be shared as needed and on demand. GRID provides the protocols, services and software development kits needed to enable flexible, controlled resource sharing on a large scale. This sharing is, necessarily, highly controlled, with resource providers and users defining clearly and carefully just what is shared, who is allowed to share, and the conditions under which sharing occurs. A GRID system is naturally highly dynamic and should be able to adapt at runtime to changes in system state as resource availability may fluctuate. Grid users are members of *virtual organizations/communities.* A virtual organization (VO) is a dynamic collection of individuals, institutions and resources sharing common goals, bundled together in order to share resources and services.

GRID technologies have evolved from ad hoc solutions, and de facto standards based on the Globus Toolkit, to Open Grid Services Architecture (OGSA) [5] which adopts Web service standards and extends services to all kind of resources (not only computing and storage). Foster et al. call service: *a (potentially transient) stateful service instance supporting reliable and secure invocation (when required), lifetime management, notification, policy management, credential management, and virtualization.* OGSA introduces two major aspects in SOA by distinguishing service factory from service instance. In other words, services are instantiated with their own dedicated resources and for a certain amount of time. These characteristics enable (i) service state management: Grid services can be either stateful or stateless; (ii) service lifetime management: Grid services can be either transient or persistent. More recently, the Web Service Resource Framework (WSRF) [6] defines uniform mechanisms for defining, inspecting, and managing stateful resources in Web/Grid services. WSRF models Grid service as an association, called a WS-Resource, between two entities: a stateless Web service, which does not have state, and stateful resources which do have state. A stateful service has an internal state that persists over multiple interactions.

2.2 Integration Related Work

Some work has already been proposed for using agents to enhance Web services or integrating MAS & SOC. For a detailed comparison between these two concepts see, for example, [11]. [12] points out some drawbacks of Web services which significantly distinguish them from agents. According to us, different kind of approaches may be distinguished in agent-Web service integration:

Distinct view of agents and Web services. Agents are able both to describe their services as Web services and to search/use Web services by using mappings between MAS standards and SOA standards [11,13,14,15]. This approach is often based on a gateway or wrapper which transforms one standard into another. As the main approach in agent standardization is the one of Foundation for Intelligent Physical Agents (FIPA), this work only considers FIPA agents and resolves relationships between SOA and FIPA standards. A particularly difficult factor in this approach is communication. The challenge consists of bridging the gap between semantically reach asynchronous based agent communications and semantically poor synchronous based Web service communications.

Uniform view of agents and Web services. Agents and Web services are the same entities. All services are Web services and they are all provided by agents (i.e., the underpinning program application is an agent-based system) [16,17].

MAS to support SOC/SOA mechanisms. This approach is not directly interested in agent service-Web service interaction but rather in the use of MAS to enhance SOAs. For example, [18] discusses the use of agents for Web services selection according to the quality of matching criteria and ratings.

MAS-based Business Process Management. Workflow or service orchestration is analogous to interaction protocol in agent communication. Both terms describe a common interaction structure that specifies a set of intermediate states in the communication process as well as the transitions between these states. The applicability of MAS to workflow enactment has been noted by [19]. More specifically, [14] makes a strict comparison between workflow (with Business Process Execution Language for Web Services) and interaction protocol (as FIPA has defined them). Conversation or service choreography is also analogous to agent conversation. Conversations are long-lived high-level interactions which need a peer-to-peer, proactive, dynamic and loosely coupled mode of interaction. Using agent conversations to enhance service exchange is an active research topic [20,21,22].

There is an increasing amount of research activity in GRID and MAS convergence taking place.[3] The use of agents for GRID was very early suggested by [3]. The authors

---

[3] See, for example, 'Agent-Based Cluster and Grid Computing' workshops, 'Smart Grid Technologies' AAMAS workshops, the Multi-Agent and Grid System journal.

specifically detail how agents can provide a useful abstraction at the Computational Grid layer and enhance resource and service discovery, negotiation, registries, etc. MAS has also been established in 2001 as a key element of the Semantic Grid [4]. And more recently, why GRID and MAS need each other has been established by [1]. The authors emphasizes the overlap in problems that GRID and MAS address but without sharing research progress in either area: *an integrated Grid/agent approach will only be achieved via a more fine-grain intertwining of the two technologies.* Using MAS principles to improve core GRID performances (e.g., directory services, scheduling, brokering services, task allocation, dynamic resource allocation and load balancing) is a very active topic in the MAS community, for example: (i) MAS-based GRID for resource management [23,24,25,26]; (ii) MAS-based GRID for VO management [27].

However, none of this work proposes a real integration of MAS and GRID. Rather, they focus on how MAS and AI techniques may enhance core GRID functionalities. Our vision of a GRID-MAS integration is not a simple interoperation of the technologies. It goes beyond a simple use of one technology to enhance the other. We aim to adopt a common approach for the integration to be able to benefit from the most relevant aspects of both GRID and MAS. This common approach is centred on the concept of service.

3   Agent-Grid Integration Language

3.1   AGIL's Concepts

This section defines progressively each AGIL's concepts coming from both GRID and MAS and integrated together in a common and relevant manner. Notice that key GRID concepts presented in this section have been established according to the OGSA or WSRF specifications. Similarly, key MAS concepts have been established by different approaches in the MAS literature [22,28] but especially the STROBE model [8,9]. As we are focussing on concepts, we adopt the most convenient terminology from these sets of specifications. AGIL's integration model is graphically presented in Figure 2 and explained in the following paragraphs:

In SOC, a *service* is an interface of a functionality (or capability) compliant with SOA standards. Figure 1 presents services we aim to formalize and their associated symbols. Stateless services are quite restrictive: they are synchronous (i.e., messages can not be buffered and do block the sender or receiver), point-to-point (i.e., used by only one user) and interact via simple one-shot interaction (i.e., request/answer). A stateless service does not establish a conversation. Instead, it returns a result from an invocation, much like a function. Stateful services required additional consideration: they are instantiated with a given set of resources. They can be persistent or transient (instantiated for a given period of time, this period may change dynamically). Transient services are instantiated by a service factory whereas persistent services are created by out-of-band mechanisms such as the initialization of a new service container. Stateful services may be multipoint (i.e., used by several users) and may interact by simple one-shot interaction or long-lived conversation. Stateful services may be synchronous or asynchronous.

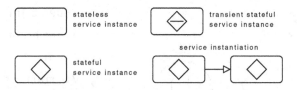

Fig. 1. Representation of key service concepts

GRID is a resource-sharing system. Grid resources are contributed by *hosts*. A host is either a *single host* (i.e., a direct association between a *computing resource* and a *storage resource*) or a *coupled host* (i.e., an aggregation of different single hosts and/or coupled hots). The sharing of these resources is implemented by the *virtualization* and *reification*[4] of these resources in a *service container*. A (Grid) service is an interface of a functionality (or capability) compliant with SOA standards. An *service instance* is included in a hosting environment in order to exist and to evolve with their own private contexts (i.e., set of resources). This is the role of the service container which is the reification of a portion of the virtualized resource available in a secure and reliable manner. A service container contains several types of services. A service may instantiate another service in the same or different service container. Each service is identified by a *handle*. Since a container is a particular kind of service, it is created either through the use of a service factory or by the direct core GRID functionality. A service container is allocated to (and created for) one and only one group of *agents*, called a *Virtual Organization* (VO). Each agent may be a *member* of several VOs. The relation between a VO and a service container is embodied by an *authorization service* which formalizes the VO-dedicated policies of service by members. The authorization service may be viewed as a MxS matrix, where M corresponds to the number of members of the VO, S to the number of currently active services, and the matrix nodes are deontic rules. These rules permit the accurate specification of the right levels for a member on a service (e.g., permissions, interdictions, restrictions etc.).[5] In order to participate in GRID, hosts and agents must hold a *X509 certificate* signed by a special authority.

An *agent* possesses both intelligent and functional abilities. These are represented respectively by the agent *brain* and *body*. The brain is composed of a set of rules and algorithms (e.g., machine learning) that give to the agent learning and reasoning skills. It also contains the agent knowledge, objectives, and mental states (e.g., Belief-Desire-Intention). The body is composed of a set of *capabilities* which correspond to the agent's capacity or ability to do something, i.e., to perform some task. These capabilities may be interfaced as Grid services in the service container that belongs to a VO an agent is a member of. In the agent's body, these capabilities may be executed in a particular conversation context called a *cognitive environment*. A cognitive environment contains several capacities. An agent may have several cognitive environments

---

[4] Resource virtualization and reification is done at the core GRID level (middleware). The rest of GRID core level mechanisms (e.g., container, authorization, etc.) are themselves described by a single unit: the Grid service.

[5] Such authorization service may be for instance, a Community Authorization Service (CAS) or Virtual Organization Membership Service (VOMS).

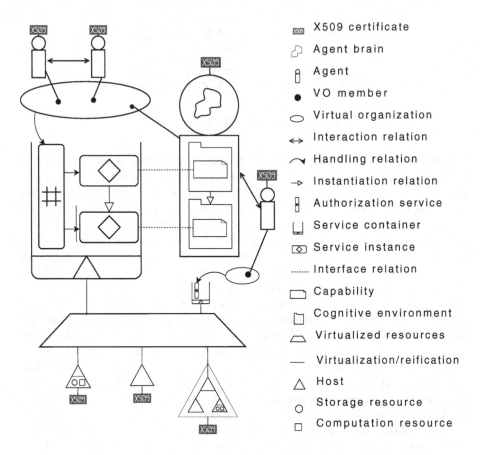

X509 certificate
Agent brain
Agent
VO member
Virtual organization
Interaction relation
Handling relation
Instantiation relation
Authorization service
Service container
Service instance
Interface relation
Capability
Cognitive environment
Virtualized resources
Virtualization/reification
Host
Storage resource
Computation resource

Fig. 2. AGIL integration model and graphical representation

which correspond to the different conversation contexts and languages it develops by
interaction with other agents. Service exchange interaction are defined when an agent
uses the service another agent provides.

In MAS, when an agent has a conversation, it dedicates a part of its state to this
conversation. It is called the conversation context [8,22]. For example, during service
exchanges, the service provider maintain a set of explicit interaction contexts, corresponding to each of its users. Conversations and their states are represented in the
STROBE model by cognitive environments. We can explore further the concept of
cognitive environments,[6] which is a relatively new, but very important, concept related to the STROBE agent and communication model [8,9,29,30]. In the STROBE
model agents are able to interpret communication messages in a given conversation
context that includes an interpreter, dedicated to the current conversation. We show how
communication enables dynamic changes in these dedicated contexts and how these

---

[6] The term environment is used here in its programming-language meaning, that is to say, a
structure that binds variables and values. It does not mean the world surrounding an agent.

interpreters can dynamically adapt their way of interpreting messages (meta-level learning by communication). Each time an agent receives a message, it selects the unique corresponding cognitive environment dedicated to the message sender in order to interpret the message. When an agent receives a message for the first time, it *instantiates* a new dedicated conversation context for this agent by creating a new one or sharing an already existing one. This instantiation mechanism is similar than the one existing in Grid services.

Having dedicated contexts and thus dedicated capabilities, is demonstrated a good means to go towards DSG. In other agent architectures, a cognitive environment may simply be viewed as a conversation context. The same concept of putting the communication contexts at the centre of the agent architecture in which it interprets messages appears also in [22], which assumes that each agent in service exchanges may separately maintain its own internal context of the conversation state.

3.2   Integration of GRID and MAS Concepts

The integration of key GRID and MAS concepts concerns five major aspects:

1. The term *agent* is used to uniformly denote Artificial Agent, Human Agent and Grid user. They are active entities involved in service exchanges. They are considered autonomous, intelligent and interactive. In particular, by viewing Grid users as agents, we may consider them as a potential artificial entities.

2. The term *VO* unifies the concept of VO in GRID and the concept of group in MAS. This is a dynamic social group (virtual or not). It is the context of service exchanges.

3. The two concepts of *service* and *capability* are linked together with a new one-to-one relation between them called the *interface relation* (represented by a dotted line in Figure 2). A Grid service is seen as the interface of a capability published in a service container and with allocated resources. An agent has a set of capabilities it may transform into Grid services available in the different VOs it is a member of. The process of 'transforming' or 'publishing' a capability into a service is called the *servicization process*.[7] When a capability is servicized, it means:

- the interfacing of this capability with SOA standards i.e., mainly WSDL/SOAP;
- the addition (possibly by using an add-service service) of this service to the VO's service container by assigning it a handle and by allocating it private resources;
- the requesting of the VO's authorization service to add an entry for this service (the agent has to decide the users' right levels);
- the publishing of the service description in the VO's registry, if it exists;
- the notification to the VO's members of the VO that a new service is available;
- etc., according to VO or service container local rules.

When an agent servicizes one of its capability into a service available for a VO it uses a set of services of this VO. Each of the previous step of the servicization process is achieved using a specific VO local service (e.g., interfacing, adding, notifications services). This servicization process is not discrete but continuous. Service and capability

---

[7] We can say that as GRID virtualizes resources and reifies them in a service container, an agent virtualizes capabilities and reifies them in a service container.

keep aligned in time one another. For example, if the capability of the agent changes, then the service changes at the same time. With this viewpoint, an agent can provide different VOs with different services. Notice also that a service is agent-specific, that means that only one agent executes (i.e., provides) the service in a container. However, it does not prevent another agent of the VO from providing the same type of service. What is important in this servicization process is that it abstracts on the kind of agent involved. Both AAs and HAs transform their capabilities in the VO's service container modulo different (graphical) interfaces. For example, an AA may servicize its capability to compute square roots (i.e., a function that receives an integer as a parameter and returns a float as result), and a HA may servicize its pattern-recognition capability (i.e., a function that receives an image as a parameter and returns a concept as result). Notice that the service and the capability lifetimes are not necessarily the same. Even if a service is transient in a service container the corresponding capability maybe persistent in the agent's body.

Remark – Grid resources are available for services (i.e., servicized capabilities) execution. The agent itself is still executed autonomously with its own resources and process (e.g., on an agent platform such as JADE).

4. The key GRID idea of service instantiation is integrated with the MAS idea of creating a dedicated conversation context. The processes are the same but viewed differently. The new conversation context contains the new capability and the service provider applies the servicization process on it in order to make available the new service instance for the service user(s). The association between the conversation context (stateful) and the including capability (stateless) is view as a WS-Resource.[8] Integrating these two instantiation mechanisms make capabilities to benefit from standardization, interoperation and allocated resources from GRID, and Grid services to benefit from a dedicated context of execution and local conversation representation from MAS.

5. Agent-agent interactions include all other kinds of interactions (Grid user-Grid service, Grid service-Grid service, agent-agent, etc.). These interactions are realized by means of asynchronous message passing between agents. There is two kind of interactions:

Direct agent-agent interaction. Messages are exchanged directly from agent to agent. These are interactions in a general sense, i.e., any interaction, standardized or ad hoc, protocol guided or not, semantically described or not, long-lived or one-shot, etc. These interactions may occur within a VO, but also outside it;

Through-service agent-agent interaction. They occur during service exchange. Messages are exchanged from agent to agent through a service. These are interactions that an agent may have with another agent, without directly communicating with the other agent but instead via the service interface this second agent offers in the VO's service container. These 'through-service interactions' occur only within a VO.

What is important in this integrated model is to consider how a service may be adapted by a service provider agent for a service user agent, in order to implement DSG. We identify four ways:

---

[8] In order to map exactly the STROBE mechanisms to OGSA and WSRF mechanisms, we should say that a new cognitive environment may be viewed as a new WS-Resource, i.e., a dedicated association between capabilities and stateful resources.

1. The service provider agent adapts the dedicated service according to its interactions with service user agent;
2. The service provider agent may offer another service to change or adapt the original service (meta-level);
3. The service provider agent may use dynamic intelligent reflection rules to change the service it is currently providing;
4. Direct agent-agent interactions may occur between the service user agent and the service provider agent and within these interactions (1) and (3) may occur in a pure ad hoc form (not via service).

3.3 Discussions and Benefits for GRID, MAS and SOC

Some AGIL advantages may be summarized:

- There is no real standard in the MAS community to describe agents' capabilities between different agents or MAS. The integration will help MAS developers in presenting and interfacing agents' capabilities, and therefore augment MAS interoperation and standardization.
- This integrated model does not restrict MAS or GRID in any way. In particular, it does not prevent direct agent-agent interactions and thus, for example, it does not prevent agents to perform tasks to one another in a purely ad hoc manner. This is important if we want the integration to be followed by numbers of MAS approaches and models; these models can keep their internal formalisms for their internal operations.
- In this integration, VO management benefits from both GRID and MAS organizational structure formalisms, e.g., Agent-Group-Role [28], CAS service, X509 certificate, etc.
- Service exchanges in this integrated model benefit from the important agent communications abilities, e.g., dealing with semantics, ability to hold a conversation, etc. The challenge in MAS of modelling conversation not by a fixed structure (interaction protocol) but by a dynamic dialogue becomes the same that dynamically composing and choreographing services in business processes as suggested by DSG.
- This integrated model subsumes a significant number of the MAS-based GRID approaches cited in section 2.2. Indeed, thanks to the reflexivity of GRID, which defines some GRID core functionalities as (meta-)Grid services (e.g., service container, auhtorization service), we may consider these core GRID services as executed also by agents. This establishes an important part of the MAS-based GRID approaches which use MAS techniques to enhance core GRID functionalities.

4 Conclusion

Identifying key factors to demonstrate the convergence of MAS and GRID models is not an easy task. We point out that the current state of GRID and MAS research activities is sufficiently mature to enable justifying the exploration of the path towards an integration of the two domains. At the core of this integration is the concept of service. The bottom-up vision of service in GRID combined with the top-down vision of service in MAS bring forth a richer concept of service, integrating both GRID and MAS properties. We put this enhanced concept of service into the perspective of Dynamic Service Generation (DSG).

In our integrated model, we consider agents exchanging services through VOs they are members of: both the service user and the service provider are considered to be agents. They may decide to make available one of their capabilities in a certain VO but not in another. The VO's service container is then used as a service publication/retrieval platform (the semantics may also be situated there). A service is executed by an agent with resources allocated by the service container. We sum-up here AGIL's two main underlying ideas:

- The representation of agent capabilities as Grid services in service containers, i.e., viewing Grid service as an 'allocated interface' of an agent capability by substituting the object-oriented kernel of Web/Grid services with and agent oriented one;
- The assimilation of the service instantiation mechanism – fundamental in GRID as it allows Grid services to be stateful and dynamic – with the dedicated cognitive environment instantiation mechanism – fundamental in STROBE as it allows one agent to dedicate to another one a conversation context.

In [31] we propose Agora, an architecture model that uses GRID to deploy collaborative ubiquitous spaces for collective intelligence. AGIL integration model is demonstrated as a key element for such an infrastructure. AGIL model is feasible considering today's state of SOC, MAS and GRID technologies. Integrating these aspects according to the guidelines given in this paper seems to us a good way to capitalize past, present and future work in order to simplify the scenarios and use fruitfully the power of distributed services, exchanged among communities of humans and artificial agents.

References

1. Foster, I., Jennings, N.R., Kesselman, C.: Brain meets brawn: why Grid and agents need each other. In: 3rd International Joint Conference on Autonomous Agents and Multiagent Systems, AAMAS 2004, New York, NY, USA, July 2004, vol. 1, pp. 8–15 (2004)
2. Huhns, M.N., Singh, M.P., Burstein, M., Decker, K., Durfee, E., Finin, T., Gasser, L., Goradia, H., Jennings, N.R., Lakkaraju, K., Nakashima, H., Parunak, V., Rosenschein, J.S., Ruvinsky, A., Sukthankar, G., Swarup, S., Sycara, K., Tambe, M., Wagner, T., Zavala, L.: Research directions for service-oriented multiagent systems. Internet Computing 9(6), 65–70 (2005)
3. Rana, O.F., Moreau, L.: Issues in building agent based computational Grids. In: 3rd Workshop of the UK Special Interest Group on Multi-Agent Systems, UKMAS 2000, Oxford, UK, pp. 1–11 (December 2000)
4. Roure, D.D., Jennings, N.R., Shadbolt, N.: Research agenda for the Semantic Grid: a future e-science infrastructure. Technical report, University of Southampton, UK (June 2001); Report commissioned for EPSRC/DTI Core e-Science Programme
5. Foster, I., Kesselman, C., Nick, J., Tuecke, S.: The physiology of the Grid: an Open Grid Services Architecture for distributed systems integration. In: Open Grid Service Infrastructure WG, Global Grid Forum, The Globus Alliance (June 2002)

6. Foster, I., Frey, J., Graham, S., Tuecke, S., Czajkowski, K., Ferguson, D.F., Leymann, F., Nally, M., Sedukhin, I., Snelling, D., Storey, T., Vambenepe, W., Weerawarana, S.: Modeling stateful resources with Web services. Whitepaper Ver. 1.1, The Globus Alliance (May 2004)

7. Singh, M.P., Huhns, M.N.: Service-Oriented Computing, Semantics, Processes, Agents. John Wiley & Sons, Chichester (2005)

8. Jonquet, C., Cerri, S.: The STROBE model: Dynamic Service Generation on the Grid. Applied Artificial Intelligence, Special issue on Learning Grid Services 19(9-10), 967–1013 (2005)

9. Jonquet, C.: Dynamic Service Generation: Agent interactions for service exchange on the Grid. PhD thesis, University Montpellier 2, Montpellier, France (November 2006)

10. Jonquet, C., Dugenie, P., Cerri, S.A.: Agent-Grid Integration Language. Multiagent and Grid Systems (2008); Accepted for publication - In press expected Number 1 vol. 4 (2008)

11. Moreau, L.: Agents for the Grid: a comparison with Web services (part 1: the transport layer). In: Bal, H.E., Lohr, K.P., Reinefeld, A. (eds.) 2nd IEEE/ACM International Symposium on Cluster Computing and the Grid, CCGRID 2002, pp. 220–228. IEEE Computer Society, Berlin (2002)

12. Huhns, M.N.: Agents as Web services. Internet Computing 6(4), 93–95 (2002)

13. Lyell, M., Rosen, L., Casagni-Simkins, M., Norris, D.: On software agents and Web services: usage and design concepts and issues. In: 1st International Workshop on Web Services and Agent Based Engineering, WSABE 2003, Melbourne, Australia (July 2003)

14. Buhler, P.A., Vidal, J.M.: Integrating agent services into BPEL4WS defined workflows. In: 4th International Workshop on Web-Oriented Software Technologies, IWWOST 2004, Munich, Germany (July 2004)

15. Greenwood, D., Calisti, M.: Engineering Web service - agent integration. In: IEEE Systems, Cybernetics and Man Conference, SMC 2004, The Hague, Netherlands, vol. 2, pp. 1918–1925. IEEE Computer Society, Los Alamitos (2004)

16. Ishikawa, F., Yoshioka, N., Tahara, Y.: Toward synthesis of Web services and mobile agents. In: 2nd International Workshop on Web Services and Agent Based Engineering, WSABE 2004, New York, NY, USA, pp. 48–55 (July 2004)

17. Peters, J.: Integration of mobile agents and Web services. In: 1st European Young Researchers Workshop on Service-Oriented Computing, YR-SOC 2005, Leicester, UK, Software Technology Research Laboratory, April 2005, pp. 53–58, De Montfort University (2005)

18. Maximilien, E.M., Singh, M.P.: Agent-based architecture for autonomic Web service selection. In: 1st International Workshop on Web Services and Agent Based Engineering, WSABE 2003, Sydney, Australia (July 2003)

19. Singh, M.P., Huhns, M.N.: Multiagent systems for workflow. Intelligent Systems in Accounting, Finance and Management 8(2), 105–117 (1999)

20. Maamar, Z., Mostéfaoui, S.K., Lahkim, M.: Web services composition using software agents and conversations. In: Benslimane, D. (ed.) Les services Web, RSTI-ISI, Lavoisier, vol. 10, pp. 49–66 (2005)

21. Hanson, J.E., Nandi, P., Levine, D.W.: Conversation-enabled Web services for agents and e-business. In: 3rd International Conference on Internet Computing, IC 2002, Las Vegas, NV, USA, June 2002, pp. 791–796 (2002)

22. Ardissono, L., Goy, A., Petrone, G.: Enabling conversations with Web services. In: 2nd International Joint Conference on Autonomous Agents and Multi-Agent Systems, AAMAS 2003, Melbourne, Australia, pp. 819–826. ACM Press, New York (2003)

23. Manola, F., Thompson, C.: Characterizing the agent Grid. Technical report 990623, Object Services and Consulting, Inc. (June 1999)

24. Cao, J., Jarvis, S.A., Saini, S., Kerbyson, D.J., Nudd, G.R.: ARMS: an Agent-based Resource Management System for Grid computing. Scientific Programming, Special issue on Grid Computing 10(2), 135–148 (2002)
25. Shen, W., Li, Y., Ghenniwa, H.H., Wang, C.: Adaptive negotiation for agent-based Grid computing. In: 1st International Agentcities Workshop on Challenges in Open Agent Environments, Bologna, Italy, July 2002, pp. 32–36 (2002)
26. Manvi, S.S., Birje, M.N., Prasad, B.: An agent-based resource allocation model for computational Grids. Multiagent and Grid Systems 1(1), 17–27 (2005)
27. Patel, J., Teacy, W.T.L., Jennings, N.R., Luck, M., Chalmers, S., Oren, N., Norman, T.J., Preece, A., Gray, P.M.D., Shercliff, G., Stockreisser, P.J., Shao, J., Gray, W.A., Fiddian, N.J., Thompson, S.: Agent-based virtual organisations for the Grid. Multiagent and Grid Systems 1(4), 237–249 (2005)
28. Ferber, J., Gutknecht, O., Michel, F.: From agents to organizations: an organizational view of multi-agent systems. In: Giorgini, P., Müller, J.P., Odell, J.J. (eds.) AOSE 2003. LNCS, vol. 2935, pp. 214–230. Springer, Heidelberg (2004)
29. Cerri, S.A.: Cognitive Environments in the STROBE model. In: Brna, P., Paiva, A., Self, J. (eds.) European Conference in Artificial Intelligence and Education, EuroAIED 1996, Lisbon, Portugal, October 1996, pp. 254–260 (1996)
30. Cerri, S.A.: Shifting the focus from control to communication: the STReam OBjects Environments model of communicating agents. In: Padget, J. (ed.) Collaboration between Human and Artificial Societies, Coordination and Agent-Based Distributed Computing. Lecture Note in Artificial Intelligence, vol. 1624, pp. 74–101. Springer, Berlin (1999)
31. Dugénie, P.: UCS, Ubiquitous Collaborative Spaces on an infrastructure of distributed resources. PhD thesis, University Montpellier 2, Montpellier, France (December 2007)

Discovering Homogenous Service Communities through Web Service Clustering

Wei Liu and Wilson Wong

School of Computer Science and Software Engineering
University of Western Australia
Crawley WA 6009
{wei,wilson}@csse.uwa.edu.au

Abstract. Contemplating the enormous success of the Web and the reluctance in taking up the web service technology, the idea of a service engine enabled service-oriented architecture seems to be more and more plausible than the traditional registry based one. Automatically clustering WSDL files on the Web into functional similar homogenous service groups can be seen as a bootstrapping step for creating a service search engine and at the same time reduce the search space for service discovery. This paper devises techniques to automatically gather, discover, and integrate features related to a set of WSDL files, and cluster them into naturally occurring groups.

1 Introduction

Web services are distributed autonomous software components that are self-describing and designed by different vendors to provide certain business functionalities to other applications through an Internet connection [1]. They are conceived to leverage existing business process creation from *tightly coupled component-based models*, to *loosely coupled* Service-Oriented Architectures (SOA). Business processes can therefore benefit from the services offered by other organisations, and are no longer limited to within the enterprise's boundary. Note that in the definition, Web services are designed to be used by other software programs automatically. However, software programs do not have any cognitive power to understand a programming interface like human programmers do. Despite more than half a decade's effort, automatically discovering web services is still considered as difficult as looking for a needle in the haystack [2]. It is widely accepted that the current SOA assumes the interactions between three types of players, namely, the *service providers* advertise their services with *service registries* and *service consumers* query the registries for providers that have matching services. Such registry-based SOA inevitably requires a semi-centralised structure, where registries become the bottleneck for scalability and robustness. In other words, if the registry (or the federation of registries) fails to perform, the service consumers and the service providers are left unconnected. Moreover, the registering of services is a static and labourious process which demands the programmer's understanding of categorisation in a domain. This is

R. Kowalczyk et al. (Eds.): SOCASE 2008, LNCS 5006, pp. 69–82, 2008.

against the open and dynamic nature of the Web. Just like the current document-centric Web where documents can be added or deleted with no central control, any Web service should be free to join and leave the service-oriented Web anytime. The registry-based SOA is fundamentally ill-fated because such a system assumes all service providers to register their new services and deregister unavailable services. In fact, some major providers have even decided to advertise their services through their human-readable web sites, rather than service-registries. For example, Google's and Amazon's Web services all have dedicated Web pages for human readers.

This paper proposed a mechanism for clustering Web Services to bootstrap a service search engine. This paper takes advantage of a document search engine (e.g. Google) that maybe unconsciously crawling web service description files (e.g. WSDL files), and using these files as seeds to start expanding the discovery of possible features in an attempt to cluster the web services into functionally similar groups. Because of the similar functionalities, we term such service clusters as *homogeneous service communities*. If the crawling and the clustering process are in continuous operation like a typical search engine does, the approach has the potential of enabling self-organisation of the Web as proposed in [3]. The proposed web service clustering approach assumes no registries, and can automatically reduce the search space of web services effectively. Therefore, it can be seen as a predecessor for Web Service Discovery. This paper gathers real service description files from the Web instead of working on hypothetical examples. The resulting clusters not only provide a useful glimpse on what services are out there, but also an insight into the types of technologies which have proliferated in this area. Theoretically, the paper introduced the use of text mining techniques to effectively separate content words from function words, and a spreading activation inspired algorithm for cluster selection. The paper is organised as follows, Section 2 discusses the proposed approach in relation to Web Service Discovery and Document Clustering. Section 3 introduces the overall architecture of the system and the detailed process of feature mining. Section 3.2 integrates the collected features using a web service cluster discovering algorithm. Section 4 demonstrated the effectiveness of the approach using existing service description files available on the Web. Section 5 conclude the paper with an outlook to future work.

2   Related Works

Web service discovery is broadly defined as *"the act of locating a machine-processable description of a web service that may have been unknown and that meets certain functional criteria"* [4]. As pointed out by [2], web service discovery mechanisms originated from the agent match-making paradigm by employing a middle or broker agent, then evolve to the various ways of querying through a standard UDDI registry or a cloud of federated UDDI registries. The discovery mechanisms also differ according to the way the web services are described. Two dominant languages are co-existing in the industry and the academia for describing web services. WSDL is popular and adopted by the industry due to its simplicity, while OWL-S (formerly DAML-S) and WSMO are well accepted by researchers as they offer much structured and detailed semantic mark-ups. Hereafter, by Web Services we mean the services described in WSDL and Semantic Web Services are for those described using either OWL-S or WSMO. The clear separation is necessary as the techniques required by these two types of languages can be quite different. According to [2] and [5], the discovery of web services in UDDI registries typically follows an Information Retrieval approach, whereas high-level match-making techniques [6] are utilised for semantic web services due to the more structured annotation of service profiles. However, semantic web services are still only available at the academic level for testing out a practical methodology like the one proposed in this paper. Instead, we opt for the more readily available format, namely, WSDL. The simplest information retrieval approach used to query a UDDI registry is the keyword-based query matched against the textual description in the UDDI catalogue and in the tModel. To address the limitation of keyword-based queries, other more sophisticated information retrieval approaches are available, such as representing service descriptions as document vectors [7] and then applying *Latent Semantic Indexing* [8] to reduce the vector space to more significant semantic concepts that characterise the web services.

The clustering of web service files is different from the traditional web service discovery problem because there are no queries to match against. However, the idea of representing a web service using document vectors is still relevant. As we will discuss in Section 3, gathering the features for a WSDL file is not as simple as collecting description documents when assuming no central UDDI registries. Another closely related area is the conventional document or web page clustering. They both involve the discovery of naturally-occurring groups of related documents (be it web pages or WSDL files). However, web service files do not usually contain sufficiently large number of words for use as index terms or features. Moreover, the small number of words present in the web service files are erratic and unreliable. Hence, conventional, detailed linguistic analysis, and statistical techniques using local corpora cannot be applied directly for web service files clustering. The use of link analysis between WSDL files to discover related web services has also been studied. In our experiments, we employed Google API's search options for discovering web page referral or citation. However, it is discovered that most of the WSDL files do not have related web pages that provide hyperlinks to them. For the few that have hyperlinks referring to them, such WSDL files are typically educational examples for teaching how to program in a service-oriented paradigm. This observation is concurred by [9].

In short, the individual existing techniques borrowed from related research areas such as information retrieval are inadequate for the purpose of discovering functionally-related web service clusters. While there is a small number of existing approaches dedicated to the discovery of web services as mentioned above, most of them remain hypothetical in nature, and have yet to be implemented and tested with real-world datasets. On that basis, we propose an integrated feature mining and clustering approach dedicated to web service clustering, which is an important predecessor to web service discovery. In the following three sections, we will discuss the proposed approach in detail, and then present some experiments using real-world WSDL files.

3   Features Mining for Web Service Files

In this paper, we propose a system that can automatically cluster a group of WSDL files obtained by querying a search engine (e.g. Google) based on the type of file (in this case, files with a `.wsdl` extension). Figure 1 illustrate the process of mining four types of features of a WSDL file, namely, the *content*, the *context*, the *service host* and the *service name*. In this system, each web service is physically represented by its corresponding WSDL file s_i. Collectively, the set of WSDL files to be considered for web service cluster discovery is represented as S. For each s_i, there are four types of features, namely, 1) the content of the Web service is characterised by the *application-specific terms* located in the WSDL file, 2) the context of the Web service is represented by the *application-specific terms* appearing in all index web pages of publicly accessible parent directories of the current directory containing the WSDL file, 3) the service host is the second- and top-level portion of the domain name (i.e. a segment of the authority part of the URI) of the host containing the WSDL file, and 4) the service name is the name of the WSDL file. As one may note, the above four features are by no means the best or the only ones available for describing a web service. However, they are the most accessible and feasible ones to use to conduct this research.

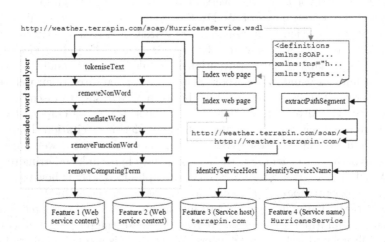

Fig. 1. Architecture of features mining. The cascaded word analyser consists of five subsequent modules for processing tokens extracted from WSDL or HTML files to extract application-specific terms as features.

These four features are collected, measured and then integrated using a web service clustering algorithm presented in Section 3.2.

3.1 Cascaded Word Analysis

Service Content and Context Extraction. To obtain the web service content, the WSDL files are first tokenised by splitting their content based on white spaces to produce a set of tokens C. The tokens $a \in C$ are essentially incomplete segments of XML elements, and can appear in various forms such as incomplete tags `<xsd:schema`. A set of heuristic rules implemented as regular expressions are utilised to remove non-word tokens. For example, tokens which exhibit signs of being part of an XML tag are removed. This process essentially reduces the set C to contain only valid words. As for the second type of features (i.e. web service context), content of web pages *"surrounding"* the WSDL file is utilised as context for describing the corresponding web service. We first identify the path segment in the hierarchical part of the URI of the WSDL file, and then recursively crawl the directories along the path. This function is depicted as the `extractPathSegment` module in Figure 1. For example, attempts are made to request for the index web pages from the directories `/portal/boulder/` and `/portal/` which are part of the URI `http://wsrp.bea.com/portal/boulder/ weather.wsdl`. All accessible index web pages are extracted and their content will undergo similar treatment as the content of the WSDL file described earlier, namely, tokenisation and removal of non-word tokens to create a second set of words X. This first step of converting XML and HTML contents into tokens, and removing non-word tokens is depicted as the `tokeniseText` and `removeNonWord` modules in Figure 1. In the second step, the morphological variants of the words in C and X are conflated through stemming and pattern matching. This step appears as the `conflateWord` module in Figure 1. Words are first reduced to their word stem. Then, we apply regular expressions to identify all word stems which are part (i.e. substring) of a longer stem. Hypothetically, we refer to such group of word stems as *word variant cluster*. Each cluster is represented by the shortest unstemmed word of a corresponding cluster member. The different variants of words are conflated into word variant clusters. The number of variants in each cluster or the cluster size is also an important piece of information that we will utilise later. Words which occur more often, including their different morphological variants, can be considered as more important features. As such, associated with each element a (i.e. normalised word appearing as shortest unstemmed word) in the new conflated sets C and X is the count of the different variants of that same word, c_a.

Content Words Recognition on the Web. After the initial sets C and X of the WSDL files have been obtained, the normalised words in these sets are analysed for their content-bearing property as shown in the module `removeFunctionWord` in Figure 1. This is step, content words are separated from Function Words using Poisson distribution. Content words are typically nouns, verbs or adjectives, and are often contrasted with function words which have little or no contribution to the meaning of texts. One of the properties of content words is that they tend to *"clump"* or to re-occur whenever they have appeared once [10]. On the other hand, the occurrence of function words tend to be independent of one another. Very often, such contrasting property can be captured through the inability of the Poisson distribution to model word occurrences in documents [11]. In other words, unlike content words, function words tend to be Poisson distributed. Following this, one way to decide if a token a is a content word or a function word is by assessing the degree of overestimation of the observed document frequency of the word a, denoted by n_a using Poisson distribution. The ratio of the estimated, $\hat{n}_a$, to the observed frequency, n_a, of word a is defined as:

$$\Lambda_a = \frac{\hat{n}_a}{n_a} \tag{1}$$

A high value of Λ_a implies an overestimation which can be used as an indicator of token a being a possible content word. In our case, any word a with Λ_a larger than the threshold Λ^T is considered as content word where

$$\Lambda^T = \begin{cases} E[\Lambda] & \text{if } (E[\Lambda] > 1) \\ 1 & \text{otherwise} \end{cases} \tag{2}$$

and $E[\Lambda]$ is the average of the observed document frequency of all tokens considered. Using Equation 2, we can identify and remove non-content words from the two sets C and X. As a result, only content words which are important in describing the associated WSDL files remain in C and X.

Application-Specific Terms Recognition through Clustering. Words such as `proxy`, `runtime`, `button` and `valign` are inevitably present in many WSDL or HTML files, and very often qualify as content words during the analysis of content-bearing property in the previous step. To obtain application-specific terms that potentially describe the functionalities of the web service, here we employ a 2-pass clustering algorithm known as the *Tree-Traversing Ant (TTA)* [12] to identify application-specific terms. This step is depicted as the last module `removeComputingTerm` in Figure 1. For the purpose of computation, we consider the structure produced by TTA as a directed acyclic graph, G. The immediate results of the TTA require human interpretation for analysing the word clusters. To facilitate the automatic selection of the desired clusters, which are groups of application-specific terms, we propose a cluster selection algorithm to complement the functioning of TTA. This cluster selection algorithm is based on the iterative propagation of *penalty weights*, ρ across the graph, and is inspired by the use of spreading activation algorithm in extending ontologies [13]. Our selection algorithm requires an oracle of sort, O, which is a predefined set of general computing terms. There are three types of vertices in the graph, namely, sinks vertices, interior vertices and a source vertex. We begin assigning weights to the sink vertices, and subsequently, to the remaining interior vertices which are the predecessors of the sink vertices. Given that V_u is the set of successor vertices of u, each vertex u is assigned a weight $0 \leq \rho_u \leq 1$,

$$\rho_u = \begin{cases} 1 & \text{if}(h_u = 0 \land u \in O) \\ 0 & \text{if}(h_u = 0 \land u \notin O) \\ |K|^{-1} \sum_{v \in V_u} \rho_v & \text{if}(h_u = 1) \\ e^{-\chi \tau_u} \sum_{v \in V_u} \rho_v & \text{if}(h_u > 1) \end{cases} \tag{3}$$

where $|K|$ is the number of sink vertices in G, h_u is the length of the longest path from vertex u to a sink (i.e. height of vertex u), χ is a decay constant, and τ_u is a measure of departure of vertex u from the origins of the weights (i.e. sink vertices). Lower values of χ result in slower decay. τ_u is only defined for interior vertices u with height $h_u > 1$ where $\tau_u = \log_{10} \left(\frac{l}{l-h_u+1} \right)$. l is the length of the longest directed path in G (i.e. length of G). As vertex u moves further away from the sink vertices, which are the origins of the weights, its height h_u increases and hence, its τ_u increases too.

3.2   Features Integration for Web Service Clustering

To discover related web services, we perform clustering using the four types of features discussed in Section 3. Similar to content words clustering during features mining, we utilise the tree-traversing ant algorithm [12] for clustering the web services. However, instead of using the word-based featureless similarity measurements with NGD and $n°W$, we introduce a new semantic relatedness measure based on the combination of the four types of features produced using our techniques described in Section 3. As pointed out before, this combination of features is necessary and critical since we do not have sufficient number of web service files for counting document frequency, and the content of the web service files is inadequate for obtaining word frequency required by vector space or probabilistic relevance models. Such issues are well recognised by some exploratory study of the WSDL files on the Web [9].

We propose a grand relatedness measure between two web services s_i and s_j, $0 \leq \Phi(s_i, s_j) \leq 1$ as:

$$\Phi(s_i, s_j) = 0.4S(C_i, C_j) + 0.3S(X_i, X_j)$$
$$+ 0.2sim(shost_i, shost_j) + 0.1sim(sname_i, sname_j) \tag{4}$$

The coefficients attached to each similarity function reflect our subjective assignment of significance of the associated features. More refined methods for combining the four features and assigning significance coefficients are possible but such discussion is beyond the scope of this paper. Both $S(C_i, C_j)$ and $S(X_i, X_j)$ are the average group similarities calculated using the formula below:

$$S(A_i, A_j) = \frac{\sum_{a \in A_i} \sum_{b \in A_j} sim(a,b)}{|A_i||A_j|} \tag{5}$$

where $sim(a, b)$ is the featureless similarity computed based on the co-occcurence of words on the Web using the *Normalised Google Distance (NGD)* [14]. $sim(a, b)$ is obtained through [12]:

$$sim(a, b) = 1 - NGD(a, b)\theta \qquad (6)$$

where $\theta = (0, 1]$ is traditionally a constant for scaling the distance NGD. However, for the sole purpose of computing S in Equation 5, we modify θ to become a variable θ_{ab} by incorporating the word variant count c associated with each conflated content word in the set C or X. The more times a word occur, either as itself or as variants in a file, the more significant that word is for describing that file. High word variant count should result in low θ_{ab} in order to produce high similarity:

$$\theta_{ab} = 0.5 \left(\frac{z_a + z_b}{z_{max}} \right) \qquad (7)$$

where $z_a = (c_a + q)^{-1} H^{-1}$ and q is a constant for adjusting the magnitude of z_a. Considering the fact that z_{max} is always achieved with the lowest word variant count $c = 1$, the value of z_{max} does not depend on word variant count

Table 1. The manually categorised dataset obtained from the Web for our experiments. There are four categories manually identified from the 22 WSDL files, namely, *"scripture"*, *"retail"*, *"weather"* and *"bioinformatics"*. These categories are used to assess the result of web service cluster discovery using the approach described in this paper. Some of the URIs and service names have been truncated with *"..."* due to space constraints.

WSDL File URI	Service Name	Service Host
http://studentmasjid.com/Quran/QuranService.wsdl	QuranService	studentmasjid.com
http://www.stgregorioschurchdc.org/wsdl/Bible.wsdl	Bible	stgregorioschurchdc.org
http://developer.ebay.com/.../ShoppingService.wsdl	ShoppingService	ebay.com
http://soap.amazon.com/schemas3/AmazonWebServices.wsdl	AmazonWebServices	amazon.com
http://www.weather.gov/forecasts/xml/.../ndfdXML.wsdl	ndfdXML	weather.gov
http://weather.terrapin.com/soap/HurricaneService.wsdl	HurricaneService	terrapin.com
http://wsrp.bea.com/portal/boulder/weather.wsdl	weather	bea.com
http://lsdis.cs.uga.edu/projects/.../WeatherFetcher.wsdl	WeatherFetcher	uga.edu
http://lsdis.cs.uga.edu/projects/.../GlobalWeather.wsdl	GlobalWeather	uga.edu
http://wwwcs.uni-paderborn.de/cs/.../GlobalWeather.wsdl	GlobalWeather	uni-paderborn.de
http://svn.codehaus.org/xfire/trunk/.../WeatherForecast.wsdl	WeatherForecast	codehaus.org
http://www.novell.com/documentation/.../weatherretriever.wsdl	weatherretriever	novell.com
http://corona.cis.temple.edu/bdhanoa/.../WeatherService.wsdl	WeatherService	temple.edu
http://services.bulport.com/weather/weather.wsdl	weather	bulport.com
http://www.govtalk.gov.uk/.../SDEP_Phase1_....wsdl	SDEP_Phase1_...	govtalk.gov.uk
http://genome.dkfz-heidelberg.de/menu/.../twofeat.wsdl	twofeat	dkfz-heidelberg.de
http://genome.dkfz-heidelberg.de/menu/.../extractfeat.wsdl	extractfeat	dkfz-heidelberg.de
http://www.cbs.dtu.dk/ws/RNAmmer/RNAmmer_1_2.wsdl	RNAmmer_1_2	dtu.dk
http://www.cbs.dtu.dk/ws/EasyGene/EasyGene_1_0.wsdl	EasyGene_1_0	dtu.dk
http://xml.nig.ac.jp/wsdl/Ensembl.wsdl	Ensembl	nig.ac.jp
http://jalapeno.health.unm.edu/svn/.../SeqServer.wsdl	SeqServer	unm.edu
http://ubio.bioinfo.cnio.es/biotools/iHOP/iHOP-SOAP.wsdl	iHOP-SOAP	cnio.es

but instead, a function of q and H. Note that the computation of z_a is based on the discrete probability distribution known as the *Zipf-Mandelbrot model* [15]. In our evaluations, q is set to 10 to obtain a more linearly distributed z_a. H is the harmonic mean defined as:

$$H = \sum_{v=1}^{c_{max}} (v + q)^{-1} \qquad (8)$$

where c_{max} is the highest word variant count in the corresponding set of feature (i.e. either C or X) across all services in set S.

4   Experiments and Results

Since there are no gold standards or readily available datasets for clustering web service files, we have resorted to manually selecting files from the top 420 query results returned by a Google search `filetype:wsdl`. Many returned results are erroneous, some are normal HTML files but happen to use `.wsdl` as the file extension. Since automatic processing does not guarantee a reliable set of test data, we manually constructed a small test set of 22 WSDL files as summarised in Table 1. To demonstrate the performance of the various aspects of features mining discussed in Section 3, we will use the outputs related to the WSDL file at the following URI `http://studentmasjid.com/Quran/QuranService.wsdl` for discussion. This service offers access to the verses and content of Islamic scriptures. The service name for this WSDL file is `QuranService` while its host is `studentmasjid.com`.

Table 2 shows a snippet of the results from the recognition of content words performed on the sets of all words in X using the Web. As we have pointed out

Table 2. A segment of the output during content-word recognition performed on the word tokens in the web service context set X for the service `QuranService`. The average overestimation Λ, or $E[\Lambda]$ is 0.99935. Based on our Equation 1, the threshold Λ^T is 1. Rows with darker shades are considered as content words since their Λ values exceed Λ^T.

word	n	n_{k-1}	$n_{k-1} = n - n_{k-1}$	$1 \geq n_{k-1}/n \geq 0$	$\hat{f}$	$\hat{f}/n \geq 1$	$\hat{n}$	$\Lambda = \hat{n}/n$
java	163,000,000	5,820,000	157,180,000	0.9643	169,035,501	1.0370	167,729,177	1.0290
developer	79,700,000	887,000	78,813,000	0.9889	80,596,983	1.0113	80,299,195	1.0075
sunnah	998,000	12,100	985,900	0.9879	1,010,249	1.0123	1,010,202	1.0122
weekly	114,000,000	290,000	113,710,000	0.9975	114,290,740	1.0026	113,692,544	0.9973
living	421,000,000	1,970,000	419,030,000	0.9953	422,979,262	1.0047	414,862,749	0.9854
document	180,000,000	6,070,000	173,930,000	0.9663	186,281,838	1.0349	184,696,189	1.0261
ramadhan	1,400,000	45,200	1,354,800	0.9677	1,446,708	1.0334	1,446,612	1.0333
hadith	1,820,000	16,900	1,803,100	0.9907	1,837,058	1.0094	1,836,903	1.0093
proxy	28,400,000	243,000	28,157,000	0.9914	28,645,097	1.0086	28,607,421	1.0073
latest	393,000,000	1,390,000	391,610,000	0.9965	394,394,934	1.0035	387,332,215	0.9856
webservice	21,100,000	27,400	21,072,600	0.9987	21,127,436	1.0013	21,106,936	1.0003
tafseer	398,000	847	397,153	0.9979	398,849	1.0021	398,842	1.0021
select	710,000,000	263,000	709,737,000	0.9996	710,263,097	1.0004	687,575,923	0.9684
recitation	1,010,000	4,330	1,005,670	0.9957	1,014,349	1.0043	1,014,302	1.0043

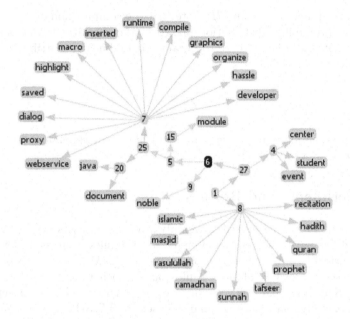

Fig. 2. The result of content word clustering on the web service context set X from the service `QuranService` using the tree-traversing ant algorithm based on featureless similarities. Visually, one can easily identify the naturally occurring content word clusters. The two biggest clusters, represented by their centroids 7 and 8, are content words describing *"general computing"* and *"Islamic studies"*.

before, there is no way of obtaining accurate single-occurrence pagecount of any word, $n_{k=1}$ required for computing the overestimation in the single-parameter Poisson, Λ. Instead, we first query search engines for multiple-occurrence pagecount of that word, $n_{k>1}$, and take the difference between n and $n_{k>1}$ as $n_{k=1}$. Some actual results obtained from the Google search engine for n, $n_{k>1}$, and $n_{k=1}$ for various words are shown in the second, third and fourth column from the left in Table 2, respectively. We utilise the wild-card search operator *"*"* provided by Google search engine for obtaining $n_{k>1}$. The fifth (i.e. $1 \geq n_{k=1}/n \geq 0$) and seventh (i.e. $\hat{f}/n \geq 1$) columns are checkpoints to ensure that the indirectly-obtained single-occurrence document frequency $n_{k=1}$, and the estimated word frequency $\hat{f}$ fall within reasonable range. The words in rows with darker shades in Table 2 are considered as content-bearing based on their overestimation Λ using the single-parameter Poisson model. The content words recognised using this method are relatively accurate. However, the actual evaluation of accuracy is beyond the scope of this paper. Figure 2 shows the results of clustering the content words in set X of service `QuranService` to identify naturally-occurring groups of words based on their genres using TTA. The two biggest clusters, represented by their centroids 7 and 8, are content words describing *"general computing"* and *"Islamic studies"*, respectively. More generally, we can see that terms which are closely related to *"general computing"* are mainly successors

Table 3. Cluster selection with $|K| = 29$. For visual inspection, we begin from the source vertex 1 which has the highest $h_1 = 6$. In both cases $\chi = 0.1$ and $\chi = 0.2$, its penalty weights ρ_1 are below $E[\rho]$, and hence, this vertex is retained. However, the next interior vertex 27 and all of its successors will be removed if $\chi = 0.1$, and otherwise if $\chi = 0.2$. All remaining vertices can be interpreted in this way.

vertex	h	τ	χ=0.2			χ=0.1		
			$e^{-\chi\tau}$	$\Sigma\rho_{successors}$	ρ	$e^{-\chi\tau}$	$\Sigma\rho_{successors}$	ρ
8	1	undef	undef	0	0	undef	0	0
9	1	undef	undef	0	0	undef	0	0
20	1	undef	undef	0	0	undef	0	0
7	1	undef	undef	3.0000	0.1034	undef	3.0000	0.1034
15	1	undef	undef	1.0000	0.0345	undef	1.0000	0.0345
4	1	undef	undef	0	0	undef	0	0
25	2	0.0792	0.9843	0.1034	0.1018	0.9921	0.1034	0.1026
5	3	0.1761	0.9654	0.1363	0.1316	0.9825	0.1371	0.1347
6	4	0.3010	0.9416	0.1316	0.1239	0.9703	0.1347	0.1307
27	5	0.4771	0.9090	0.1239	0.1126	0.9534	0.1307	0.1246
1 (source)	6	0.7782	0.8559	0.1126	0.0964	0.9251	0.1246	0.1153
E[ρ]					0.11760646			0.118648824

of the vertex 27. Table 3 shows part of the results during content word cluster selection to identify application-specific terms. For this purpose, we rely only on a small oracle containing the words $O = \{runtime, webservice,$
$developer, module, data\}$. Using $\chi = 0.1$, the penalty weight of vertex 27 is $\rho_{27} = 0.1246$. Since $\rho_{27} > E[\rho]$, vertex 27 is removed and the deletion is propagated to all successors. As a result, the final feature set X (i.e. web service context) is $\{tafseer, sunnah, recitation, ramadhan, quran, prophet,$
$masjid, hadith, islamic\}$. Finally, Figure 3 shows the web service clustering during our initial experiments using 22 actual WSDL files obtained from the Web. These files are manually categorised beforehand into four categories, namely, "scriptures", "retail", "weather" and "bioinformatics" for inspection. Next, clustering is performed on the 22 web services using the TTA algorithm with our new semantic relatedness measurement defined in Equation 4 based on the four types of features. The clustering process produces a directed acyclic graph with a single source vertex 1. For readability, dotted lines were drawn to highlight the naturally-occurring groups of web services discovered using TTA as shown in Figure 3. Each web service, appearing as a sink vertex is labeled using its shost/sname information. From the figure, we can observe that all the 22 web services are correctly assigned to their naturally-occurring groups except for nig.ac.jp/Ensembl. These groups are considered as web service clusters. The most common ancestor, appearing as an interior vertex between the web service files related to weather information is vertex 9. Similarly, the most common ancestors for services related to religious scriptures, bioinformatics, and retails are vertices 4, 5 and 3 respectively. These most common ancestors are regarded as the centroids of the corresponding web service clusters. In addition, there is an interesting trend within each of the clusters which is worth mentioning. From Figure 3, one is able to notice the further internal groupings of web services within each larger category, and such groupings are motivated by

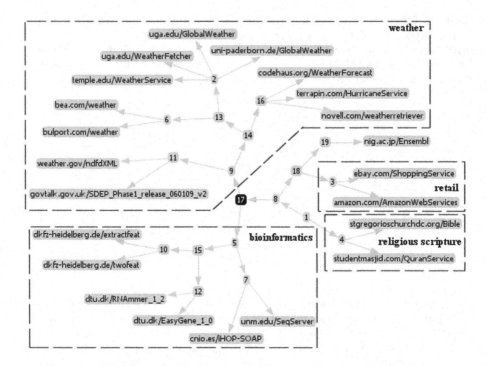

Fig. 3. The resulting web service clusters after clustering is performed using TTA based on the four types of features on the 22 web service files. The clustering produces a directed acyclic graph with a single source vertex 1. For readability, dotted shapes were drawn to highlight the naturally-occurring groups of web services which emerged during clustering. The web services are labeled in the format `shost/sname`.

the composition of the grand relatedness measure Φ. For example, in the bioinformatics service main cluster, the web services are further partitioned based on their hosts. The services which are successors to vertices 10 and 12 belong to the host `dkfz-heidelberg.de` and `dtu.dk`, respectively. In the weather service main cluster, the successors to vertex 11 are services belonging to governmental hosts `govtalk.gov.uk` and `weather.gov`. Such trend is obviously motivated by the contribution of the similarity between service hosts during the computation of Φ.

All in all, we have demonstrated the feasibility of discovering functionally-related web services through the mining of the four types of features, and clustering of web services using these features. Such approach is not only a good starting point for the development of practical service search engines, but is also capable of assisting in the design of fault-tolerant systems. The web service clusters discovered using our approach provide systems with access to redundant services in the case of a failure. Moreover, referring back to the previous paragraph, options for service redundancy do not only exist at the functional level, but also at the physical host level. In this regard, the service consumers can simply opt for functionally-similar services regardless of hosts, or can be selective in terms of the providers.

5 Conclusion and Future Work

Clustering web services into functional similar groups can greatly reduce the search space of a service discovery task. Therefore, it can be seen as a predecessor of web service discovery or an important functionality provided by future service engines. However, very few research has looked into this area. This paper proposed mining different types of features of a web service and use these features for web service clustering. To realise the proposed techniques, difficult issues such as differentiating content words from function words, and obtaining word frequencies from the Web are resolved. A spreading activation based automatic cluster selection algorithm is also implemented. The contributions of this paper extend beyond the web service community where service discovery and redundancy are important issues. The proposed approach and the output which follows are potentially useful to the text mining research community for discovering emergent semantics. Experiments and results have confirmed the feasibility and effectiveness of this automatic service clustering approach. More work is planed to evaluate this approach using more service files and potentially more feature types. The refinement of the process of features mining, especially the various modules in the cascaded word analyser, may be necessary with larger dataset. The current measurement of the grand relatedness was derived in an ad-hoc manner, and reformulation or refinement may be necessary in search of mathematical justifications.

Acknowledgement

This research was supported by the Australian Endeavour International Postgraduate Research Scholarship, and the DEST Australia-China Fund CH050103.

References

1. Peltz, C.: Web services orchestration - a review of emerging technologies, tools and standards. Technical report, Hewlett Packard, Co. (2003)
2. Garofalakis, J., Panagis, Y., Sakkopoulos, E., Tsakalidis, A.: Web service discovery mechanisms: Looking for a needle in a haystack? In: International Workshop on Web Engineering, Hypermedia Development and Web Engineering Principles and Techniques: Put them in use, in conjunction with ACM Hypertext, Santa Cruz (2004)
3. Liu, W.: Trustworthy service selection and composition - reducing the entropy of service-oriented web. In: 3rd International IEEE Conference on Industrial Informatics, Perth, Australia (2005)

4. Booth, D., Haas, H., McCabe, F., Newcomer, E.M.C., Ferris, C., Orchard, D.: Web services architecture. Technical report, W3C WG Note (2004), http://www.w3.org/TR/ws-arch/

5. Garofalakis, J., Panagis, Y., Sakkopoulos, E., Tsakalidis, A.: Contemporary web service discovery mechanisms. Journal of Web Engineering 5(3), 265–290 (2006)

6. Klusch, M., Fries, B., Sycara, K.: Automated semantic web service discovery with owls-mx. In: Proceedings of the fifth international joint conference on Autonomous agents and multiagent systems, Hakodate, Japan, pp. 915–922 (2006)

7. Sajjanhar, A., Hou, J., Zhang, Y.: Algorithm for web service matching. In: Yu, J.X., Lin, X., Lu, H., Zhang, Y. (eds.) APWeb 2004. LNCS, vol. 3007, pp. 665–670. Springer, Heidelberg (2004)

8. Berry, M.W., Dumais, S.T., O'Brien, G.W.: Using linear algebra for intelligent information retrieval. SIAM Review 37(4), 573–595 (1995)

9. Li, Y., Liu, Y., Zhang, L., Li, G., Xie, B., Sun, J.: An exploratory study of web services on the internet. In: 2007 IEEE International Conference on Web Services (ICWS) (2007)

10. Manning, C., Schutze, H.: Foundations of statistical natural language processing. MIT Press, Cambridge (1999)

11. Church, K., Gale, W.: Inverse document frequency (idf): A measure of deviations from poisson. In: Proceedings of the ACL 3rd Workshop on Very Large Corpora (1995)

12. Wong, W., Liu, W., Bennamoun, M.: Tree-traversing ant algorithm for term clustering based on featureless similarities. Journal on Data Mining and Knowledge Discovery 15(3), 349–381 (2007)

13. Liu, W., Weichselbraun, A., Scharl, A., Chang, E.: Semi-automatic ontology extension using spreading activation. Journal of Universal Knowledge Management (1), 50–58 (2005)

14. Cilibrasi, R., Vitanyi, P.: The google similarity distance. IEEE Transactions on Knowledge and Data Engineering 19(3), 370–383 (2007)

15. Mandelbrot, B.: Information theory and psycholinguistics: A theory of word frequencies. MIT Press, MA (1967)

Collaborative Learning Agents
Supporting Service Network Management

W. Mulder[1,2], G.R. Meijer [1], and P.W. Adriaans[2]

[1] LogicaCMG, Prof. Keesomlaan 14, 1180 AD Amstelveen, Netherlands
{Wico.Mulder,Geleyn,Meijer}@logicacmg.com
[2] University of Amsterdam, Kruislaan 419 Matrix 1, 1098 VA Amsterdam
adriaans@science.uva.nl

Abstract. Service oriented systems need to be maintained to keep the requested level of service. This is challenge in large grid- and saas based networks that are managed by numerous entities. This paper is about supporting multi agent systems that operate in the network and support its management by learning actual structures from life observed logging data. We focus on a collaborative grammar induction mechanism in which agents share local models in order to retrieve a model of the structure of the total service network. We studied the performance of groups of agents while varying the size and degree of communication. We motivate the application of the mechanism in the domain of service oriented system and show the results of experiments using a distributed agent-based monitoring system. We promote further research in the overlapping scientific disciplines of multi agent systems and machine learning in the application domain of service oriented systems.

1 Introduction

Service networks are networks of computer systems that are used to deliver end-user applications in a dynamic and personalized way. They are highly scalable, heterogeneous, and operate in agile, demand-driven environments. Service networks are interesting from business as well as technical perspectives; we have the business of cooperative industrial organizations providing federated services towards consumers, and we have the technical ICT infrastructures that provide, support and enable these business services in a dynamic and personalized way. These infrastructures consist of hardware and software, are based on service oriented architectures and often consist of web-services that work together in various end-user applications.

Services networks need to be maintained to keep the requested level of service. This is a particular challenge in the case of distributed networks in which the nodes are maintained by separate entities, such as in a grid based network[1]. Large service networks can be complex in terms of dimensions, interactions or level of heterogeneity. We study the operational management of such networks, and focus on maintenance support by means of autonomous software agents [12]. While operating in the

---

[1] http://www.nessi-europe.com

R. Kowalczyk et al. (Eds.): SOCASE 2008, LNCS 5006, pp. 83–92, 2008.
© Springer-Verlag Berlin Heidelberg 2008

service network itself, the task of these agents is to provide information about the status of that network and the services that are provided by it. In order to deal with the complexity of the network, the agents are designed to be adaptive, and share information with each other. A constraint often encountered in service networks is that its components only have access to local parts of the network, which is often combined with communication constraints due to security reasons or limited physical bandwidth. We believe that the dynamic and heterogeneous aspects of the environment in which the agents operate force them to collaborate in their learning and information provisioning tasks. In this way, the agents form a network themselves, providing a robust and redundant way of information provisioning and management support. Figure 1 shows the situation in which an agent network analyses the status of a service network and support the responsible network managers in their task.

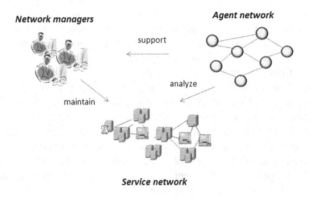

Fig. 1. Service network management support

In previous work [9] we talked about Collaborative Information Services in multi domain networks. In this paper we focus on the collaborative learning mechanism of these agents. We look at the application scenario in which the agents support network managers and system administrators to obtain the actual status and structure of a complex service network. The agents look at different sets of provenance[2] data and try to induce a grammars and structures in it. This is done by means of DFA learning, as explained in the next section where we propose a distributed DFA modeling approach.

The field of grammar induction is a well known area that has been studied from many perspectives during the last decades [6]. In a grammar induction process, a learning algorithm is used to obtain a grammar that should explain the structure of a given set of data. This grammar represents a model of the dataset. The aim is to learn

---

[2] Wikipedia: *Provenance* is the origin or source from which something comes, and the history of subsequent owners. Provenance information of some data the documentation of the process that led to the data [3]. It can be generated from the static information available in original workflow specification together with the runtime details obtained by tracing the execution of the workflow.

from sample data (usually a list of words) an unknown grammar which explains this data. The model is also used to verify whether unknown samples follow the rules of the grammar.

A Deterministic Finite Automaton (DFA) is a common algorithm used to classify structures (languages) and represent grammars in the form of graphs 0. A DFA can be seen as a model which captures the underlying rules of a system, from observations of its behavior or appearance. These observations are often represented by strings that are labeled "accepted" or "rejected". Every string in the dataset is represented as a path in the graph. Figure 2 shows an example of a DFA-tree for the strings *abcd* and *abcbcd*. By merging or clustering data using heuristics the algorithm learns to represent the data in a more structured way.

Since creating some DFA that is consistent with training data is trivial, it is usual to add two further constraints, that the DFA should generalize to unseen test data and that the challenge is to find the smallest DFA that is consistent with the training set. For the latter, we use MDL (minimum description length) as a criterion.

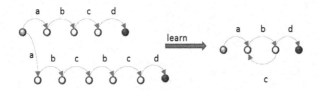

Fig. 2. Example of a unfolded DFA (left) and a folded (learned) DFA (right)

In our experiments, we use DFA as a common grammar induction method for learning individual as well as collaborative (global) models. We take samples of provenance datasets and define a set agents that learn local topology structures. The agents observe data and communicate the induced structures with each other. The goal is that each agent learns a model of the dataset as a whole.

The rest of the paper is organized as follows: Section 2 discusses our agent model and the strategies used for communication and learning. In section 3 we show the results of our first experiments. Section **Fout! Verwijzingsbron niet gevonden** contains discussion and ideas for future work, and section 5 ends with the conclusions.

2 Introduction

Our agents are designed to learn grammar in a collaborative way. We used groups of agents of different sizes, in which each agent analyzes a part of a given dataset. They observe a part of the dataset, learning a grammar from it and communicate the results with the other agents. This is done by means of two separated models per agent and a series of chosen strategies, which will be explained below.

Each agent keeps two models of the dataset; the first model, called the *individual model* reflects the structure of the observed dataset; the second model, called the *collaborative model* or *global model,* reflects the structure of the dataset as a whole. Both models are in the form of a DFA. By means of these two models per agent, we intend to have a clear separation between its local information and its shared information. Taking into account the constraints of the application domain, we cannot always communicate the original samples of the individual datasets. Therefore we have chosen to communicate the models and generate new samples from them at the moment of their arrival at the receiving agent. Figure 3 shows a schematic picture of a set of agents and their models. The arrows indicate the flow of information.

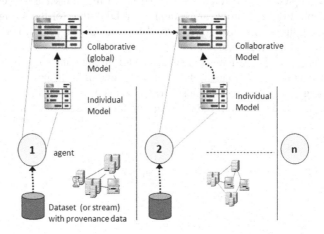

Fig. 3. N agents analyzing (workflow) data. Each agent observes data from its own local environment. The vertical lines denote the constraint that the agents do not (always) have access to the datasets of each other.

We designed our collaborative learning mechanism to work with four types of strategies: an *individual learning* strategy, a *collaborative learning* strategy, a *dispatch* strategy and an *acceptance* strategy.

The individual learning strategy reflects the way that an agents learns its individual model. In our case this is grammar induction using DFA learning. The collaborative learning strategy describes the way the collaborative model is maintained. In our experiments we combine generated samples from the individual model and incoming models from other agents and use them to build a DFA tree.

The communication process of the agents is characterized by a *dispatch strategy* and an *acceptance strategy*. Agents share their collaborative model with each other in the form of messages, called DFA-hypotheses. The dispatch strategy defines *to whom*, and *when* this communication takes place as well as the information content, the *what*, or simply the *hypothesis content*. When receiving these hypotheses, an agent uses its *acceptance strategy* to judge whether it should accept the incoming hypotheses and how they are merged with their own collaborative model.

While observing the local data sets, the agents update both, their individual model as well as their collaborative model. First, after taking a number of samples, the individual model is trained. Then data for the collaborative model is assembled from both, the individual model as well as models from other agents. The updated collaborative model is then shared again with the other agents.

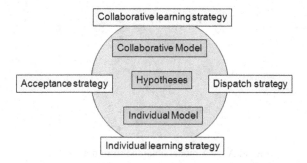

Fig. 4. An agent's internal models and strategies

Sharing takes place by communicating the model, and merging it at its arrival with the existing model. In our experiments we simply merge by building a new model from a number of generated samples from both, the previous model and the incoming one. The goal is that each agent has a good model of the total dataset.

Note that each agent has its own collaborative model, but mechanism allows these models to be slightly different (like in Plato's theory of Forms). It is common to share information between agents using blackboards [1][17]. In our mechanism the agents create and share their own collaborative model which can be regarded as an implicit, redundant, distributed blackboard.

3 Experiments

We used small datasets containing strings that represent. workflow execution orders of web-services situated in two isolated environments of a service network. An example of such a situation is shown in figure 5.

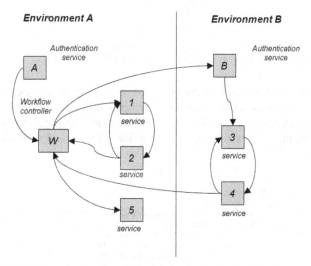

Fig. 5. Webservices and workflow orders in two isolated environments

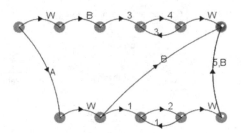

Fig. 6. A DFA model obtained from both agents

We looked at the models of the two agents, one observing data from environment A, the other observing data from environment B. An example of a learned DFA model is shown in figure 6.

We compared the collaborative models with models obtained by a single agent observing the data from both environments. Except from extra end-states, they were found to be similar.

In order to study our collaborative learning approach in detail, we carried out a series of experiments in which agents are allowed to take randomly a number of samples from a shared dataset. The agents learn individual DFA structures and share their collaborative models in order to model the total dataset.

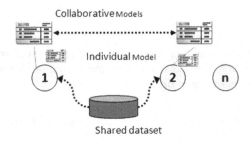

Fig. 7. Multiple agents observing a shared dataset

We studied the performance of the group while varying the number of agents (**n**), number of learning steps (**t**), the number of samples taken per learning step (**m**) and the number of samples generated from a particular model during a merge process (**s**). In each experiment we took the same web-service scenario and used a dataset of 20 different samples.

The structure of an experiment is given below.

```
Experiment (n,t,m,s):
    n agents, for each agent:
        Start with new individual model and collaborative model
        Repeat for t learning steps
            take m samples from the training set
            add unique samples to the individual model
            learn DFA from individual model samples
            generate m samples from the individual DFA
```

```
                    add generated samples, if unique, to the collaborative model
                    generate s samples from each incoming hypothesis
                    add generated samples, if unique, to the collaborative model
                    learn DFA from collaborative model
                    send DFA as hypothesis to other agents
                until t learning steps
            take whole trainingset
            verify DFA of collaborative model of each agent, determine mean score
            sum mean scores and take mean and std.
        End experiment
```

We developed an agent framework[3] allowing us to control the experiments and study the behavior of the agent network for different values of n,m,t, and s.

The collaborative model of each agent is stored in a result-database allowing us to validate and verify the performance of the agents. The collaborative model of each agent is used to classify the samples of the whole dataset. The mean number of samples classified as valid to the grammar, is defined as the score per agent. The mean score of a set of agents indicates the score of the particular experiment.

For a single agent, varying the value of m, the results are shown in fig 8. The figure indicates that for this particular dataset, the number of samples to learn a complete model for a single agent is roughly the number of samples in the dataset, which is 20.

The variance in the score strongly depends on the structure of the individual samples: during a particular experiment, samples are taken randomly and generic samples might give a valid classification result for other samples as well.

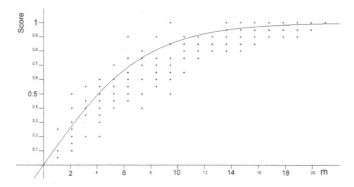

Fig. 8. Score-graph for n=1, t=1, s=0, m=1..20, the fitted logistic curve has a=1, b=0.28

The growth of the score behaves exponentially with the increase of m and saturizes to the value of 1. The steepness of this growth reflects the learning performance of the agent network for this particular dataset. We fitted the score as a function of m to a logistic curve[4], where the values 1 and 2 are used to scale the offset, parameter a is

---

[3] Existing frameworks, such as e.g. Jade, are under investigation of using instead.

[4] The logistic function has applications in areas of population statistics and biology. An example can be found in Rasch-modeling theory [11] where the probability of responses is modeled using as person and item parameters. In our model, in similar ways, the score of the agent is a mean of individual scores which in their turn are based on accepting individual samples.

taken to be 1, and parameter b fitted as an indicator of the steepness of the learning behavior.

$$f(x) = -1 + \frac{2}{1 + ae^{-bx}}$$

For a network of two and four of agents, we varied m, s and t. Figure 9 shows the results of two agents (n=2). The fitted value of b increases with the number of samples during a merge (s) and the number of learning steps (t). This means that the score climbs faster to 1 or, in other words, the agents learn from each other's examples.

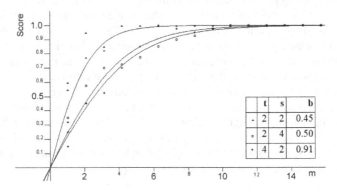

Fig. 9. Score graph of 2 agents

Figure 10 shows the same, but then for a network of 4 agents.

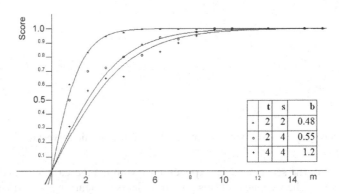

Fig. 10. Score graph of 4 agents

4 Discussion and Future Work

The area of distributed learning recognizes that in many cases agents cannot simply solve problems individually and need to combine their models. Shen and Lesser [10] have studied Distributed Bayesian Networks, where agents generate local solutions based on their own data and then transmit these high level solutions to other agents.

Network management using agents is being studied in the fields of grid computing and provenance management. Forestiero [4] studies ant-based resource management and discovery where agents copy and move resource metadata among grid hosts. Feng [3] describes a decentralized provenance recording and collection mechanism in which mobile agents collect information about jobs in workflow executions.

In our research we focus less on the actual distribution of provenance data. Instead of proper metadata management, we study the learning behavior of a set of communicating agents having collaborative models.

Since DFAs are common and fundamental in the area of unsupervised learning, we used this algorithm in our learning tasks. We do not try to improve DFA algorithms themselves, but focus on the collaborative learning behavior of the agents. Since the agents take into account hypotheses from other agents whilst they are learning themselves, our mechanism can be regarded as a 'distributed on-line learning mechanism'.

In our design and implementation of the prototype we took in account that the strategies, which are currently rather simple and straightforward, can be replaced by other, more sophisticated ones. We intend to improve the communication strategy including decision and dissemination algorithms in which receiving agents can actively ask for information as well.

We want to improve the mechanism of merging incoming hypothesis-DFAs with the model of a receiving agent using genetic algorithms; rather than searching for the best hypotheses to be taken into account for merging models, mutation and recombination of the best currently known may lead to evolutionary learning behaviors. An agent that receives hypotheses can learn to choose optimal (listening) actions to achieve its goal. As a feedback, an agent might provide a reward or penalty in reaction to an incoming and accepted hypothesis. This could be done by comparing the fitness of the collaborative model before and after the included hypothesis. On the level of meta-learning, we think of using a kind of feedback to the individual learning process; the learning process of the agent itself might be affected by incoming hypotheses from other agents.

Last but not least, for the work described in this paper, we used simple example data. Since our motivations for this research are based on expected needs and constraints in the application domain of service networks, we plan to apply our methods in this area dealing with real workflow execution data.

5 Conclusions

In this paper we talked the application domain of service oriented systems and dynamic infrastructures, in which these agents can support the operational and technical maintenance. We focused on the provision of workflow topology information obtained from provenance datasets , explained the architectural model of our agents as well as our approach of distributed DFA learning.

We have presented an approach for distributed grammar induction using collaborative agents. We showed the results of experiments in which a agents learned individual DFA models from local datasets and shared these models in order to obtain a DFA model that represents the total dataset.

We showed the results of our experiments where different groups of agents learned structures from a shared dataset, while sharing their intermediate results with each other. We analyzed the learning behavior of the agent network, and showed the relationship of its steepness with the number of agents and level of communication.

We suggested a number of improvements and promoted further research in combined fields of machine learning, multi agent systems and service oriented systems.

References

[1] Cicchello, O., Kremer, S.C.: Inducing grammars from sparse data sets: A survey of algorithms and results. Journal of Machine Learning Research 4, 603–632 (2003)

[2] Corkill, D.D.: Collaborating software: Blackboard and multi-agent systems & the future. In: Proceedings of the International Lisp Conference, New York (October 2003) (Invited presentation)

[3] Feng, Y., Cai, W.: Provenance Provisioning in Mobile Agent-based Distributed Job Workflow Execution. In: 7th International Conference on Computational Science (ICCS 2007), Beijing, China, May 27-30 (accepted, 2007)

[4] Forestiero, A., Mastroianni, C., Spezzano, G.: A decentralized ant-inspired approach for resource management and discovery in grids. Journal of Multiagent and Grid systems, 43–63 (2007)

[5] Gannon, D.: Programming the Grid: Distributed Software Components, P2P and Grid Web Services for Scientific Applications. Journal of Cluster computing, Special issue on Grid Computing (July 2002)

[6] de la Higuera, C.: A bibliographical study of grammatical inference. Pattern Recognition 38, 1332–1348 (2005)

[7] Jiang, Y.C., Yi, P., Zhang, Z.Y., Zhong, Y.P.: Constructing agents blackboard communication architecture based on graph theory. Computer Standards & Interfaces 27, 285–301 (2005)

[8] Mitchell, T.: Machine Learning. McGraw-Hill, New York (1997)

[9] Mulder, W., Meijer, G.R.: Distributed information services supporting collaborative network management. In: IFIP International Federation for Information Processing, Network-Centric Collaboration and supporting frameworks, Proceedings PROVE 2006, vol. 224, pp. 491–498. Springer, Heidelberg (2006)

[10] Shen, J., Lesser, V.: Communication Management Using Abstraction in Distributed Bayesian Networks, in Autonomous Agents and Multiagent Systems. In: AAMAS 2006, Proceedings of the fifth international joint conference, pp. 622–629 (2006) ISBN:1-59593-303-4

[11] http://en.wikipedia.org/wiki/Rasch_model,
http://en.wikipedia.org/wiki/Logistic_function

[12] Wooldridge, M.: Multi Agent Systems, Introduction to Multi Agent Systems. John Wiley and Sons, Chichester (2002)

A Multi-Agent Architecture for NATO Network Enabled Capabilities: Enabling Semantic Interoperability in Dynamic Environments (NC3A RD-2376)

Brenda J. Powers

NATO Consultation, Command and Control (C3) Agency
Oude Waalsdorperweg 61, 2597 AK The Hague, Netherlands
brenda.powers@nc3a.nato.int

Abstract. The use of Autonomous agents in conjunction with semantic web technologies such as Extensible Markup Language (XML), Resource Description Framework (RDF), Web Services and Ontologies will enhance the effectiveness of NATO Network Enabled Capabilities (NNEC) operations. These semantic web components are the enablers for realizing the benefits that agents can bring towards achieving interoperability at the semantic level. This paper presents the Semantic Interoperability Collaborative Multi-Agent Architecture (SI-CoMAr), which is proposed in conjunction with Semantic web components, to address the issue of enabling semantic interoperability in dynamic environments such as those supported by the NNEC concept of operations.

Keywords: Semantic Interoperability, NATO Network Enabled Capabilities (NNEC), Agent Technology, Ontology, Command and Control (C^2).

1 Introduction

The marriage of agent technology and semantic web concepts with respect to facilitating semantic interoperability in the NATO Network Enabled Capabilities (NNEC) concept of network-centric warfare, and in Command and Control (C^2) operations in particular, is promising. The emergence of Semantic Web technologies and specifications, and standards, that have been developed to provide a formal description of concepts, terms and relationships for specific knowledge domains, are the optimal enablers for Autonomous Agents to understand, acquire and integrate information more efficiently and intelligently. Semantic web components such as Ontologies and Web Services are key to realizing the benefits that autonomous agents can bring towards achieving interoperability at the semantic level in a net-centric environment.

In dynamically changing environments, there is a requirement for proactive components which can perceive and adapt to the situation at hand. Cognitive autonomous agents that proactively act on behalf of users, to collect, analyze and fuse data, discover services, monitor data assets, and facilitate information and knowledge management can be instrumental in providing relevant information, which enables the achievement of a greater degree of situational awareness for the entire network.

R. Kowalczyk et al. (Eds.): SOCASE 2008, LNCS 5006, pp. 93–103, 2008.
© Springer-Verlag Berlin Heidelberg 2008

The emergence of proactive, agent-based, adaptive software greatly improves situational awareness at the operational level of war by facilitating decision support for both planning and execution in an NNEC environment. Autonomous Intelligent agents are a solution for proactively handling many tasks. Agents that continuously monitor the events in the operational environment can assist in providing information for operational users such as analysts and planners to conduct threat analysis, terrain analysis, asset scheduling and tracking, route planning, logistical planning, Search and Rescue operations, force protection planning, and coordination with NATO, National and Civilian forces.

This paper will present a multi-agent architecture for supporting semantic interoperability in net-centric environments, which can be used for development of software agents that utilize ontologies and web services to support human-computer collaboration, in the area of decision support for C^2 operations conducted in the NNEC environment.

The remainder of this paper is organized as follows: Section 2 gives an overview of the Semantic Interoperability (SI) Project, Section 3 presents the description of the Semantic Interoperability Collaborative Multi-Agent Architecture (SI-CoMAr), developed as part of the SI project. Section 4 introduces several use cases for use of SI-CoMAr in conjunction with semantic web technologies, in support of Decision Support, and Maritime Situational Awareness, operations in an NNEC environment. Section 5 summarizes the conclusions and Section 6 describes future work.

2 Semantic Interoperability Project

The Semantic Interoperability Project is a research effort begun in 2005. The initial task was to design and build a concept demonstrator of the Semantic Interoperability Mediation Services (SIMS). SIMS is focused on the conversion, aggregation, and routing of information and consists of a set of modules that stand alone as services; SI Search, Inference Service (IS), Knowledge Store (KS), NATO Metadata Registry and Repository (NMRR), and Service Discovery Service (SDS). All these services have been designed with standard specifications in mind and with a Service Oriented Architecture (SOA) approach, providing both SOAP and ReST-based interfaces [1].

The SI Search communicates with third party applications by supporting ReST, SOAP and SPARQL queries.

The Inference Service (IS) provides support for discovering, retrieving and reasoning with the different service provider vocabularies. It includes the reasoning module (Pellet Reasoner), and query translation functionality. IS provides a SOAP-based web service interface with a well defined WSDL.

The Knowledge Store (KS) maintains the information collected by the SI Search for a consumer query. The IS makes use of the KS to reason about the information and extend it with implicit knowledge derived from a set of rules (ontologies).

The NATO Metadata Registry and Repository (NMRR) project aims to capture all the necessary information and perform the groundwork required for procurement of a NATO Metadata Registry. Through registration in the NMRR, a wide range of specifications and services available in the NNEC environment will be visible and accessible.

The Service Discovery Service (SDS) provides a dynamic and automatic mechanism to detect and keep an up-to-date repository of available services. This repository will change in near real-time as service availability changes. Any consumer service that wants to know which services are available at any given time, will use the SDS interface to retrieve the endpoints of such services. The SI Search makes use of this functionality acting on behalf of the SI Search consumer.

The Semantic Interoperability Collaborative Multi-agent architecture (SI-CoMAr) has been developed as part of the SI project and will be used to realize an agent component which can utilize, the NMRR and SDS to discover web services, the semantic search, KS, IS, ontologies, and the RDF data stores which have been developed by the SI project.

3 Semantic Interoperability Collaborative Multi-Agent Architecture (SI-CoMAr)

This section describes the Semantic Interoperability Collaborative Multi-agent architecture (SI-CoMAr), which is proposed to be used in conjunction with Web Services technology and other Semantic Web components to facilitate Interoperability at the Semantic level in support of C^2 operations in NNEC environments.

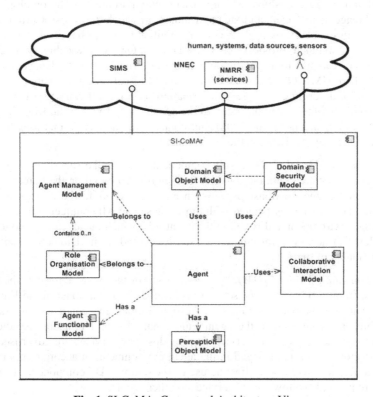

Fig. 1. SI-CoMAr Conceptual Architecture View

The SI-CoMAr also supports the concept of developing a society of Cognitive, Autonomous software entities, communicating, acting, learning and making pro-active decisions to support a human-computer partnership.

In Figure 1, the SI-CoMAr conceptual architecture view shows that agents exist in a Multi-agent system (MAS) component which resides in an NNEC environment and has an interface to the SIMS component, services via the NMRR, users, data sources, and sensors.

The components in the SI-CoMAr architecture are described as follows:

Agent Management Model (AMM): This component is designed in accordance with the FIPA specification on Agent Management, which supports the concept of an Agent Platform (AP) or world in which the agents reside. An interesting point relative to NNEC environments is that agents can live in a world that is distributed over many host computers. Within the MAS, the agents are managed by the Agent Management Model, which handles agent creation and deletion and migration of agents between platforms. The AMM contains the Role Organization Model, which contains the Agent IDentifier (AID) and Description obtained at agent creation.

Domain Object Model (DOM): The DOM is the knowledge representation or description of the environment in which the agents are situated. The model consists of Ontologies, which represent the set of concepts, relationships, and objects and their attributes and the relationships, which exist in the environment. The ontologies are used by the agents to understand the environment in which they exist. Currently, the SIMS Domain Ontology by itself is not a monolithic structure but a set of small ontologies that, combined, provide the overall Domain Ontology vocabulary. Each ontology file corresponds to a specific domain system, (i.e., Imagery Management and Reporting Tool (IMART), Networked Interoperable Real-time Information Services (NIRIS) etc..), describing the system information model vocabulary [2]. Thus, additional domain ontologies can be added to the agents' world view. The Domain Security Model is dependent on the DOM and contains the access permissions for data sources and services available in the environment.

Domain Security Model (DSM): This component contains the security policies and access rights associated with the domain. It is configured based on the domain ontology and contains the relationships between agent roles within the domain and the information access rules associated with those roles. The AID uniquely represents an agent in the environment and is tied to the agent role, which makes it possible to control the level of access to information or services listed in the Directory Facilitator (DF) component of the AP.

Role Organization Model (ROM): This component of the AMM describes the types of Agents, (monitoring, analysis, service discovery, etc..), that exist in the Environment. A role is comprised of functional responsibilities and characteristics (attributes), which are contained in the agent Functional Model. The agent role dictates what type of functionality and services it provides. The services are advertised in a global directory which is managed in the agent environment, in accordance with the FIPA specification, which specifies the use of the optional DF component of the AP for the provision of 'yellow pages' services to other agents.

Perception Object Model (POM): This component defines the agent's beliefs, desires and intentions, which are used in the dynamic reasoning process. The Agent Functional Model defines the methods an agent will use to perform its tasks, which are selected in accordance with the agent's perception. Thus, the ROM in conjunction with the POM and the Agent Functional Model (AFM) define who the agent is in the Environment, what his beliefs or perceptions are, and the functions he uses to perform the tasks necessary to achieve his goals.

Agent Functional Model (AFM): Methods used to carry out Plans, contained in the POM, are located in the AFM. Based on Role, the AFM model contains the methods that agents use to carry out intentions. Plan methods can be combined into a sequence of actions the agent performs to execute the Plan. Other methods include those used to achieve tasks such as managing access to services, managing ontologies and services, monitoring information sources, and gathering information from heterogeneous sources.

Collaborative Interaction Model (CIM): Agents communicate with each other via messages formatted using an Agent Communication Language (ACL). The Collaborative Interaction Model (CIM) is designed in accordance with the FIPA Agent Communication language (ACL) specification and uses an implementation of the FIPA Agent Message Transport Service (MTS), which is defined in the FIPA Agent MTS Specification [3]. The CIM handles the coordination of messages exchanged. Agents discover and collaborate with other agents in the MAS based on the tasks to be performed. The part of the CIM which directs collaboration between agents, utilizes a yellow pages and agent management service to locate agents. It also references a set of ontologies that support message content in the messages exchanged during communication between agents in the environment.

4 SI-CoMAr Application To NNEC

The emergence of proactive, agent-based, adaptive software greatly improves situational awareness at the operational level of war by facilitating decision support for both planning and execution in an NNEC environment. Autonomous Intelligent agents are a solution for proactively handling many tasks. Agents can discover web services registered in the NMRR to assist in providing information for operational users such as analysts and planners to conduct threat analysis, terrain analysis, asset scheduling and tracking, route planning, logistical planning, Search and Rescue operations, and coordination with NATO, National and Civilian forces.

A networked environment promotes and facilitates the acquisition, sharing, and application of knowledge from various heterogeneous databases. Intelligent agents offer the potential for large volumes of data to be collected, intelligently integrated via use of ontologies and rule sets, and displayed without overloading users with too much irrelevant information. Consequently, traversal, retrieval and intelligent processing of data in the environment, leads to information at a higher level of aggregation. The result is more efficient situational awareness in command centers all over the battle space [4].

Given these facts, the next two sections explore a few domain specific use cases which describe how agents can be used effectively with semantic web components, to facilitate situational awareness in support of military decision-makers in the areas of C^2 and Maritime operations.

4.1   Command and Control: Decision Support

The emergence of intelligent, agent-based, adaptive software in conjunction with semantic web technologies, greatly improves military capabilities at the operational level by providing mechanisms to assist in decision support for both planning and execution. Agents can also assist decision makers in threat assessment and intelligent information integration in support of providing better situational awareness.

Decision-makers are required to assess and solve a variety of problems as quickly as possible, at times sometimes, without adequate resources. The incorporation of agent technology into C^2 applications offers great benefit in the form of human-computer collaboration; established to assist decision-makers in carrying out their mission related activities. If agents are to effectively assist human decision-makers in accomplishing their C^2 mission related activities, they must possess enough autonomy so that they can behave in a pro-active manner in order to be of maximum benefit in a human computer partnership [5]. While this is true, the abilities of human decision makers in the areas of conceptualization, abstraction and creativity [6] far surpass their agent counterparts, whose strengths lie in computational speed, parallelism, accuracy, and data assimilation and management.

Continuous sensing of the battlespace; a fundamental reordering of information configuration and distribution; and, integrated reasoning are needed to support operational decision makers. Use of adaptive intelligent agents could be employed to analyze and reason about information stored in semantically encoded formats such as RDF and Web Ontology Language (OWL), and monitor the battlespace in support of threat assessment. So why use agents for this? In dynamically changing environments where situations change there is a requirement to be able to recognize and adapt to the requirements of the new situation. Some examples are given in the following section.

C^2 Use Cases

Consider the following use cases in which SI-CoMAr Analysis, Service Discovery, Monitoring and Information Management agents could be deployed:

***Analyze and Correlate Intelligence Information*:** Given the availability of data sources, Analysis agents could correlate information much faster than a human being, because they do not get tired or require sleep. Agents can assist in learning trends, creating profiles based on historical patterns, and using prior knowledge attained to analyze and correlate various sources of intelligence information. For example, while deciding what is of interest must ultimately be done by a human analyst, Analysis agents could filter events with interesting characteristics among the many events which might appear in a dataset. Agents also have the potential to assist humans, in reasoning about complex networks of relationships.

***Perform Service Discovery*:** Service Discovery agents act on behalf of NNEC users, to locate services in the NMRR, which satisfy the user's requirements. For example, users in the Crisis Planning Domain need to have all the information on the weather in a certain location in order to plan a Search and Rescue operation. Perhaps, this particular rescue operation involves the use of a helicopter and a small ship, thus weather services providing information on both tides and wind velocity would need to be discovered by the agent. The list of available services providing this information could be further narrowed down based on what the agent has learned in the past about the services. For example, if the agent has learned that Service X has not performed as advertized, or Service Y always provides outdated information, then these services would not be selected by the agent in the future.

Monitor Suspicious Track Activity: Monitoring agents are responsible for watching specified data assets, sensors, and information sources in the NNEC environment, they monitor and generate alerts when Air or Maritime Contacts of Interest (COI) come within a specified radius of a protected port or renegade aircrafts fly off course into protected airspace.

Monitor Networked Interoperable Real-time Information Services (NIRIS) Interface: The Networked Interoperable Real-time Information Services (NIRIS) system displays real time maritime, ground, air tactical and theatre missile defense data received from control reporting centers. NIRIS allows a single common air picture, gathered from various radars linked to control and reporting centers (CRCs), NATO Airborne Early Warning (NAEW) aircraft and other tracks received at command and control entities anywhere in Allied Command Operations (ACO). There are three cases that the monitoring agents will consider. (1) Agents monitor the NIRIS Tactical Data Link (TDL) streams located at the CAOC, to detect if they go down. If the link goes down, Information Management agents will modify all NIRIS Network Port Manager (NPM) configuration files. (2) Agents monitor NIRIS to determine if a new radar has been deployed or an existing radar has been relocated. In either case the Information Management agents will modify the Site Configuration files at all locations where NIRIS is deployed. (3) Monitoring agents detect that NIRIS is down and recognize that, because access to a certain data asset has been lost, certain web services listed in the NMRR are no longer available for consumption. The agents notify the SDS and the appropriate adjustments are made to the NMRR.

Monitor Call For Fire (CFF) messages: Agents check CFF messages and issue alerts when the rules of engagement are violated and enemy or friendly units were directly or indirectly targeted.

4.2 Maritime Situational Awareness

NATO's Maritime Situational Awareness (MSA) capability can be described as the effective understanding of anything associated with the global maritime environment that could impact the security, safety, economy or environment of the Alliance.

The Maritime environment includes all areas and things relating to the oceans, seas, bays, estuaries, waterways, coastal regions, and their corresponding airspace [7].

MSA involves the collection, fusion and dissemination of enormous quantities of data drawn from government agencies, military, and commercial entities. MSA consists of two key components, information and intelligence. Observing the maritime situation involves collecting, fusing and monitoring open source information. The large number of vessels and the high amount of information associated with them makes this task extremely challenging.

MSA Use Cases

There are a variety of use cases that Agents could satisfy in the area of MSA, the primary two are discussed as follows:

***Information Discovery and Integration*:** This process involves the extraction of not only information from government and military agencies, but a vast amount of open source data from web sites such as the Paris MOU, Lloyd's Register Fairplay, Lloyd's SeaSearcher, classification societies, and AISLive. Information such as, registration data for vessels, port information, electronic ship certificates, detention lists and more, is all used by MSA operational users. It must be intelligently integrated in such a way so as to provide knowledge for the operational users. Table 1 provides

Table 1. MSA Commercial Data Sources

Data Source	Description
AISLive	Automatic Identification System (AIS) is a shipboard broadcast system that acts like a transponder, operating in the Very High Frequency (VHF) maritime band. Information including the ship's identity, type, position, course, speed, navigational status and other safety related information, is broadcast.
Classification societies	Organizations that establish and apply technical standards in relation to the design, construction and survey of marine related facilities including ships and offshore structures. Information is available as open source on the internet.
Lloyd's Register Fairplay	A commercially available data source that provides reference information on vessels and shipping companies. This company issues the International Maritime Organization (IMO) number on behalf of the IMO. It has details of over 8,300 ports and terminals, casualty data, fixtures, vessel detentions, photographs, an electronic news archive going back over ten years and real-time vessel movements.
Lloyd's Marine Intelligence Unit	Provides merchant vessel information such as, arrival, bound for and departure details from over 4,000 ports, characteristics for each vessel including tonnages, dimensions, capacities and full engine details, 10-year casualty history for every vessel including serious and non-serious incidents, and 10-year detention history for each vessel including details of past inspections and indication as to whether the inspection led to formal detention.
Paris MOU open source information	The Paris Memorandum of Understanding (MOU) consists of 27 participating maritime Administrations and covers the waters of the European coastal States and the North Atlantic basin from North America to Europe. The open source information contains information on banned ships and current and past detention lists.

some insight into the type of information available. For further details concerning MSA commercial data sources, please see the original document from which the information was obtained [8].

Agents could provide intelligent information integration and monitoring of the AIS based maritime picture and flag vessels that show any signs of suspicious behavior for closer investigation. The objective in using agent technology is to proactively maximize synergy of contributions from maritime intelligence and information agencies to produce an accurate Common Operational Picture (COP) situation in order to detect and take action on illegal activity.

Threat Assessment: Merchant shipping information is monitored on the NATO unclassified network by the Commercial Shipping Cell at Northwood and the Data Operations Gathering (DOG) team in Naples. They monitor the merchant shipping (AIS-based) picture and available communications media to trigger and guide specific data gathering, data verification, assessment and alerting activities in relation to merchant shipping, where alerts can be configured for vessels worthy of further investigation or action. These indicators may be derived from unclassified data, such as, vessels deviating from declared routes, vessels exhibiting some types of suspicious behavior such as operating below certain speed thresholds, or at anchor outside normal anchorage areas, contradictory or unknown vessel identity, or a record of safety-related detentions which might indicate operators or owners are susceptible to criminal behavior.

Agents could assist operational users during the Threat Assessment process, by monitoring and collecting available vessel information from various sources. Analysis agents can integrate large sets of collected data, reason about, and alert the operator of any inconsistencies and anomalies detected. Information such as prior record of illegal activities and detentions, AIS data, and AIS-based vessel data from commercial sources including Lloyd's Fairplay, can be retrieved by collection agents via the SI Search, and fused in order to determine potential threats.

5 Conclusions

This technical note has introduced the Semantic Interoperability Collaborative Multi-agent architecture (SI-CoMAr), which is proposed to be used in conjunction with Web Services technology and other Semantic Web components, developed as part of the SI Project, to facilitate Interoperability at the Semantic level in support of C^2 operations in NNEC environments. SI-CoMAr supports the concept of developing a society of Cognitive, Autonomous software entities, communicating, acting, learning and making pro-active decisions to satisfy user specified requests.

This paper has explained the benefits of using agent technology in conjunction with semantic web technologies, such as web services, RDF stores, and ontologies in support of software applications which support C^2 and Intelligence operations in net centric environments.

Specific use cases in which SI-CoMAr can be combined with web services and ontologies to benefit operational users in the areas of Threat Assessment, Decision

Support, and Maritime Situational Awareness were presented. Clearly, agents are better utilized to proactively perform certain tasks in dynamic environments, such as discovering information and services, monitoring data assets, and performing timely, fusion and integration of information. While this is true, it is equally as clear that there must be a human-computer partnership in which the Human remains in the loop.

6 Future Work

We are currently working on implementing a prototype which will use the JADE framework to realize the SI-CoMAr in support of human computer collaboration in the area of Threat Assessment, Decision Support and MSA. The IS, KS, ontologies, RDF data stores, NMRR, SDS, and other semantic web based components created as part of the SI Project will be incorporated into this effort.

The JADE framework provides a Web Services Integration Gateway (WSIG) which enables interoperation between JADE agents and Web services. Use of the WISIG and extension of it, using the Java API for XML Registries (JAXR) which provides a uniform and standard Java API for accessing different kinds of XML Registries, is under investigation.

The use of the JADEX reasoning engine, which implements the BDI model and can be used with the JADE framework, is under investigation.

Acknowledgements

The author thanks the North Atlantic Treaty Organization C4I Standardization & Architecture branch of Allied Command Transformation (ACT), for their support of this work. The work was performed by a NATO Civilian employee and grants NATO royalty-free permission to reproduce all or part of the contribution and to authorize others to do so for NATO purposes.

References

1. Clarke, D., et al.: Semantic Interoperability Executive Summary. NATO Consultation, Command and Control Agency Technical Paper (2007)
2. Rodrigues-Herola, V., et al.: NNEC Semantic Interoperability. NATO Consultation, Command and Control Agency Technical Note 1234 (2005)
3. Foundation for Intelligent Physical Agents (FIPA), FIPA Agent Message Transport Service Specification, FIPA67 (2000)
4. Ceruti, M., Powers, B., Wilcox, D.: Knowledge Management for Command and Control. Command and Control. Research and Technology Symposium. San Diego (June 2004)
5. Ceruti, M., Powers, B.: Intelligent Agents for FORCEnet: Greater Dependability in Network-Centric Warfare. In: IEEE International Conference on Dependable Systems and Networks (July 2004)

6. Powers, B.: Adaptive Intelligent Agents in Informative Environments: Human-Computer Collaboration in Command and Control Application Environments. In: Space and Naval Warfare Systems Center, San Diego Biennial Review, vol. TD 3117, pp. 203–207 (2001)
7. Pohl, J., Chapman, A., Pohl, K.: Computer-Aided Design Systems for the 21st Century: Some Design Guidelines. In: 5th International Conference on Design and Decision-Support Systems for Architecture and Urban Planning (August 2000)
8. Grainger, J.: CC MAR Naples MSA Processes (2006), http://tide.act.nato.int/mediawiki/index.php/MDA_Overview
9. Grainger, J.: Maritime Situational Awareness - Merchant Shipping Information Categories. NATO Consultation, Command and Control Agency Technical Note 1244 (2006)

An Agent for Asymmetric Process Mediation in Open Environments*

Roman Vaculín[1], Roman Neruda[1], and Katia Sycara[2]

[1] Institute of Computer Science, Academy of Sciences of the Czech Republic
{vaculin,roman}@cs.cas.cz
[2] The Robotics Institute, Carnegie Mellon University
katia@cs.cmu.edu

Abstract. The ability to deal with incompatibilities of service requesters and providers is a critical factor for achieving interoperability in dynamic open environments. We propose a Process Mediation Agent (PMA) as a solution to the process mediation problem in situations when the requester does not want to reveal its process model completely for privacy reasons. The PMA automatically resolves encountered incompatibilities by generating mappings between processes of the requester and the provider and applies them for the runtime translations. In the PMA algorithms we combine the AI planing and semantic reasoning with recovery techniques and the discovery of appropriate external data mediators.

1 Introduction

The research in the Web Services field focuses on the goal of enabling and facilitating smooth interoperability of heterogeneous distributed software components. Devising WS standards such as SOAP, WSDL, BPEL4WS together with additional semantic layers such as WSDL-S [1], OWL-S [2] and WSMO [3] presents one step towards this goal. Additionally, a lot of effort is being invested in automated techniques for service discovery and composition [4]. However, this is not enough for achieving the goal, at least for two reasons.

First, service providers and requesters do not typically share the same data models, interaction protocols and oftentimes even not the basic standards for WS specification. As a result, we have to deal with incompatibilities on different levels among service providers and requesters [5]. Second, due to the dynamic nature of the Internet and rapid, unpredictable changes of business needs, also the existing web services change very often. The ability to adapt to changing environments is therefore crucial.

For these reasons, in open environments it is very unlikely that a dynamically discovered service provider will be able to interoperate directly with a service requester even if we assume that both partners share the same domain ontologies. Various types of middle agents [6] present a possible solution for bridging the gap between service requesters and providers with fixed incompatible interaction protocols (process models)

---

* Research partially supported by the "Information Society" project 1ET100300517 and the Czech Ministry of Education project ME08095.

R. Kowalczyk et al. (Eds.): SOCASE 2008, LNCS 5006, pp. 104–117, 2008.

and possibly incompatible data models. We believe, that software agents technology — employing techniques such as reasoning and planning combined with approaches like dynamic discovery and recovery from failure — is the best choice available today that can offer an alternative to dealing with problems of incompatibilities and adaptivity manually.

In this paper we focus on the process models mediation problem of service requesters and providers operating in open dynamic environments where both the requester and the provider interact according to specified process models that are fixed and that might be incompatible. In open environments service providers are motivated to publish their process model in order to maximize the usage rate and thus the profit. On the contrary, clients typically concern about the privacy aspects and therefore do not wish to fully reveal their process models. In our view, such an asymmetric setting is more realistic in open environments than a symmetric scenario which would be relevant in closed or semi-closed environments of corporate networks [5].

The solution we propose consists of developing a process mediation middle agent (PMA) which automatically resolves the incompatibilities by generating appropriate mappings between processes of the requester and the provider and applies them for the runtime translations. The PMA uses external services (external data mediators) to deal with data incompatibilities and missing pieces of data together with planning techniques to find the appropriate mappings by combing available external data mediators. Since the PMA operates in dynamic open environments, it needs to interact with an appropriate discovery service. In our case, the PMA uses an external OWL-S Matchmaker service [7] to take care of the discovery of data mediators. Finally, the PMA integrates comprehensive recovery mechanisms based on compensation, backtracking and recovery operations such as *replace*, *retry* and *replacyByEquivalent* [8].

The main contributions of the paper are the following. We describe the problem of the asymmetric process mediation in open environments that we believe is becoming important especially if we consider the increasing importance of mobile and pervasive computing (Section 2). Next, we introduce a possible solution in the form of the PMA agent (Section 3) that combines the AI planing and semantic reasoning (Section 4) with the discovery of appropriate external data mediators (Section 5) and suitable recovery techniques (Section 6).

2   Mediation Problem

When requesters and providers use fixed, incompatible communication protocols interoperability can be achieved by applying a *process mediation agent* which resolves all incompatibilities, generates appropriate mappings between different processes and translates messages exchanged during run-time. As we mentioned in the previous section, the PMA has access to the complete process model of the provider, while for privacy reasons it can see only the current request message (possibly semantically annotated) received from the requester.

The PMA has to address various types of incompatibilities on different levels. On the *data level*, services may be using different formats to encode elementary data or data can be represented in incompatible data structures using different syntactic and lexical representations. Possible ontology mismatches also belong in this category. On the *process level*, messages can be exchanged in different orderings, some pieces of information which are required by one process may be missing in the other one, control flows can be encoded in different ways, one service call of the requester can be realized by several calls of the provider, information might need to be reused, etc. In [5] more details about possible incompatibilities are given. Conceptually, we distinguish *data mediators* that are responsible for resolving data level mismatches from *process mediators* that are responsible for resolving process level mismatches. Typically, when trying to achieve interoperability, process mediators and data mediators are closely related. A natural way is to use data mediators within the process mediation component to resolve "lower" level mismatches that were identified during the process mediation.

Because the problem of process mediation is complex and extensive, we focus on the process level mismatches while we address the data mediation only in a very limited way. For details on the data mediation see, e.g., [9,10]. We assume that *data mediators* have the form of external services which can be discovered and used by the PMA. Furthermore, in our system data mediators can have a form of converters that are built-in to the PMA system. Currently, built-in converters support basic type conversions such as up-casting and down-casting based on reasoning about types of inputs and outputs. By up-casting or down-casting we mean a conversion of an instance of some ontology class to a more generic or more specific class respectively.

We assume that both the requester and the provider behave according to specified process models and that both process models are expressed explicitly using OWL-S ontologies [2]. In OWL-S, the elementary unit of process models is an atomic process, which represents one indivisible operation that the client can perform by sending a particular message to the service and receiving a corresponding response. Processes are specified by means of their inputs, outputs, preconditions, and effects (IOPEs). Types of inputs and outputs are defined as concepts in an ontology or as simple XSD data-types. Processes can be combined into composite processes by using control constructs such as sequence, any-order, choice, if-then-else, split, loops, etc.

Finally, we assume that both process models share the same domain ontology and target conceptually the same problem.

3   The PMA Overview and Architecture

The problem of process mediation can be seen as finding an appropriate mapping between requester's and provider's process models. Assuming that the requester starts to execute its process model, ideally, we want to show that for each step[1] of the requester the provider (with some possible help of intermediate translations) can satisfy the requester's requirements (i.e., providing required outputs and effects) while respecting its own process model. Since we are restricted by the fact that neither of process models is known in advance the PMA needs to find the mappings only during the runtime. Also, only the current sequence of executed steps is known to the PMA, while the overall requester's process model is not revealed to the PMA for privacy reasons. Thus,

---

[1] In the following text the word *step* stands for an atomic process executed by the requester. If we refer to the provider's atomic processes, we mention it explicitly.

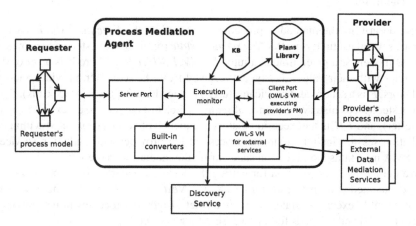

Fig. 1. Mediation of process models by the Process Mediation Agent (PMA): problem setting and the PMA architecture

the PMA, after receiving the requester's request, needs to decide if the request can be mapped into some provider's atomic process (or a sequence of processes) given the current provider's execution state, available data and data mediators. If such a mapping is found, the PMA executes it and returns the results to the provider. The PMA tries to remember and reuse information that it gained from previous interactions with requesters and a discovery service. Only when no historical information is available, the PMA explores the search space to find an appropriate mapping by simulating the execution of the provider's process model. Since all the interactions with external web services can fail, and also many choices made by the PMA can lead to failure, the PMA tries to recover from failures and possibly use backtracking if more mappings are available for a given execution state.

Figure 1 shows an architecture of the PMA. The *server port* is used for interactions with the requester and the *client port* for interactions with the provider. The *client port* uses the OWL-S Virtual Machine (OVM) [11] to interact with the provider. The OVM is a generic OWL-S processor for execution of OWL-S services with built-in advanced features such as support for recovery and execution monitoring [8]. Another instance of the OVM is used to execute external data mediation services if necessary. The *Execution Monitor* is the central part of the PMA. It executes the mediation algorithm and links all the other components together. Specifically, the *Execution Monitor* maintains the execution state and stores information received from the requester and provider in a *Knowledge Base*. Furthermore, the *Execution Monitor* interacts with the *Discovery Service* when new data mediators need to be found. The *Plans Library* is used to store reconciliation plans (defined in Section 3.1) that were used successfully in the previous mediation sessions for the purposes of future mediations. Also information about discovered data mediators is cached in the *Plans Library*.

3.1 Definitions

Before describing the mediation procedure executed by the *Execution Monitor* we introduce several useful terms. PMA's *execution state* for given requester's and provider's process models at a given time is a tuple $S = \langle V, F, RH, EH \rangle$, where V is a set of data (variables with their values and types) received from the requester, provider and data mediators, F is a set of valid predicates (e.g., produced as effects of service calls), RH is a requester's steps history (sequence of requests, i.e., atomic processes, executed by the requester), and EH is an execution history (sequence of services executed within one mediation session including atomic process of the provider and mediation services). The PMA maintains the execution state during the mediation in the *KB*.

We define the *Next* function for getting the set of provider's atomic processes that can be executed at a given execution state, i.e. $Next : \mathscr{S} \rightarrow \mathbb{P}(\mathscr{PA})$, where $\mathscr{S}$ is the set of possible execution states, $\mathscr{PA}$ is the set of atomic processes in the provider's process model and $\mathbb{P}$ stands for a power set of a given set.

We say that the request r can be satisfied by the atomic process p in the execution state S and write $satisfied(r, p, S)$ if all inputs of p are either provided by r or are available in S, all preconditions of p are satisfied in S and all outputs and effects required by r are produced by p or are available in S.

Internally, we represent mappings between provider's and requester's processes as *reconciliation plans*. The *reconciliation plan* for a request r and execution state S specifies which actions in what order need to be executed to perform the translations between r and provider's process in the state S. Furthermore, the reconciliation plan can also represent what queries need to be asked to the discovery service to find new data mediators so that the plan can be executed.

The *reconciliation plan M* for a request r and execution state S is a tuple $M = \langle P, Q \rangle$, where P is a partially ordered plan consisting of data mediator processes, atomic processes from $\mathscr{PA}$, and unbound *abstract processes*, and Q is a possibly empty set of *query templates*.

An *abstract process* in the plan is a place-holder for another plan which needs to be specified later. Abstract process is associated with some query template in Q which can be used for finding the plan that will be used in the place of the abstract process.

A query template q is a tuple $q = \langle I, O, E, S \rangle$ where I is a set of available inputs, O is a set of required outputs, E is a set of required effects and S is the execution state. We use query templates to represent discovery requirements for the discovery service (details in Section 5).

The reconciliation plan $M = \langle P, Q \rangle$ is called *executable* if there are no abstract processes in P, and Q is empty. Otherwise, the plan is called *unbound*.

3.2 Top-Level Mediation Procedure

The logic of the mediation procedure is the following. For each requester's request the PMA calls the *processRequest* procedure until requester or provider finishes successfully or the execution fails. Algorithm 1 shows high-level steps of the *processRequest* procedure. In this procedure, the least time consuming mediation options are considered first, and only if they fail other possibilities are considered.

Algorithm 1. Procedure *processRequest*

1. Receive a requester's atomic process call *requesterCall* via the *server port*
2. Store inputs of *requesterCall* in the state S and required results in the *KB*
3. Find the best available mediation actions:

   3.1 **if** $A = \{a \,|\, a \in Next(S) \wedge satisfied(requesterCall, a, S)\} \neq \emptyset$ **then**

      // *Exact match: no data mediation needed*

      – **foreach** $a \in A$ **do**

         • **if** $execute(a, S)$ fails **then** $localRecover(a, S)$ and **continue**

         • **else return success**

   3.2 **if** exists reconciliation plan P in the *Plans Library* for *requesterCall* and S

      // *Reusing existing plan*

      – **if** $execute(P, S)$ fails **then** $localRecover(P, S)$ and **continue**

      – **else return success**

   3.3 **if** $reconciliationPlans = reconcileRequestCall(requesterCall, S) \neq \emptyset$ **then**

      // *Planning was used to find the mapping*

      – **foreach** executable reconciliation plan $P \in reconciliationPlans$ **do**

         • **if** $execute(P, S)$ fails **then** $localRecover(P, S)$ and **continue**

         • **else** store P in the *Plans Library* and **return success**

      – **foreach** unbound reconciliation plan $P \in reconciliationPlans$ **do**

      // *If no directly executable plan found, use discovery service*

         • **if** $P' = bindPlan(P)$ succeeds **then**

            ∗ **if** $execute(P')$ fails **then** $localRecover(P', S)$ and **continue**

            ∗ **else** store P' in the *Plans Library* and **return success**

   3.4 **if** $globalRecover(S)$ fails **then** // *Nothing worked, undo and pick a another branch*

      – **return failed**

After the PMA receives the request it first tries to match it to some provider's atomic process available in the given execution state (step 3.1).

If no such process exists or execution of all of them fails, reconciliation plans from the *Plans Library* are considered (step 3.2).

As the next step (3.3), the planning algorithm is used (*reconcileRequestCall* procedure, for details see Section 4) that tries to find new reconciliation plans by combining known data mediators or proposing queries to the discovery service which would allow to bridge gaps (mismatches) identified during the reconciliation. If some executable plan is found and successfully executed it is stored in the *Plans Library*. Otherwise, if only unbound plans are found, the discovery service needs to be used to bind the plan in the *bindPlan* procedure (see details in Section 5).

Finally, if none of the mediation alternatives succeeds, there still might be a chance that in some previous phase a wrong branch in the provider's process model was taken which does not allow the mediation to be finished while another branch might work. The *globalRecover* procedure in step 3.4 tries to deal with such a situation (see Section 6 for details on recovery).

The $execute(P, S)$ procedure executes the reconciliation plan P by executing every action $a \in P$. If a is a provider's process the client port is used, while if a is a data mediator the OVM for external services is used.

Algorithm 2. *reconcileRequestCall*(r, S), r a request call, S execution state

---

1. **Initialize:**
   - **foreach** $i \in r.inputs$ **do** add *(Available i.name i.type)* *(Value i.name i.value)* to S
   - **foreach** $o \in r.outputs$ **do** add *(RequesterGoal (Available o.name o.type))* to S
   - **foreach** *effect* $\in r.effects$ **do** add *(RequesterGoal effect)* to S
   - *plans* $= \emptyset$
2. **foreach** $p \in Next(S)$ **do** *reconcileAtomicProcess*$(r, p, S, plans, \emptyset, maxPlanLength)$
3. **return** *plans*

We left out some details in the algorithm, in particular the ordering of matched processes and plans in steps 3.1 and 3.3. The processes with an *exact* match with respect to the *requestCall* are considered first, followed by processes with *plug-in* and *subsumes* match (for details on types of match known from the discovery literature see e.g. [7]).

4 Finding Reconciliation Plans

The purpose of the reconciliation procedure for a given requester's request r and an execution state S is to find possible reconciliation plans that can be used to reconcile the mismatches between the request r and the current state of the provider's process model. The reconciliation procedure uses a planning algorithm that tries to combine known data mediators to find necessary transformations. If no combination of known data mediators can be found, the reconciliation procedure produces an unbound plan associated with query templates which can be used later to discover new data mediators and to bind the plan.

In principle, the reconciliation procedure transforms missing pieces of information (inputs, outputs, preconditions and effects) into goals that need to be satisfied. Then a classical backward chaining planning algorithm is employed with data mediators and providers atomic processes used as planning operators. In order to guarantee timely termination of the planning algorithm, the maximal length of the plan is constrained externally. During the planning, the operations are only simulated (services are not executed) with respect to the initial execution state S.

Algorithm 2 takes care of the planner initialization (step 1) and starting the reconciliation for every provider's atomic process available in S (step 2). The core of the reconciliation algorithm is performed by the *reconcileAtomicProcess* procedure displayed in Algorithm 3. This procedure is trying to find a plan for reconciliation of request r and process p. First, it must guarantee that all inputs and preconditions of p are available (step 2), and afterwards that all outputs and effects required by r were produced (step 3). In both cases the *solveGoals* procedure implementing a backward chaining algorithm is tried first to achieve missing goals by means of known data mediators. If *solveGoals* does not succeed (i.e., some goals cannot be satisfied), the query template q for the discovery service is suggested that would allow a discovery of new data mediators needed for finishing the plan.

An important remark is related to plans ranking. Generated reconciliation plans are ranked depending on their quality and length. The quality of a plan is derived from the

---

Algorithm 3. *reconcileAtomicProcess*$(r, p, S, plans, plan, maxPlanLength)$, r a request call, p provider's atomic process, S execution state, *plans* a set of all plans, *plan* current plan, *maxPlanLength* maximal allowed length of generated plans

1. **if** $|plan| \geq maxPlanLength$ **then return**
2. **Reconcile inputs and preconditions of** p**:**
   - **if** ($\forall i \in p.inputs$ available in S) and ($\forall prec \in p.preconditions$ satisfied in S) **then**
     - simulate p (add outputs and effects of p to S); add p to *plan*
   - **else**
     - *Goals*= transform missing inputs of p and unsatisfied preconditions of p to goals
       // e.g., missing toCode input => (Goal(Available toCode AirportToCode))
     - **if** *solveGoals*$(Goals, S, plan, maxPlanLength)$ **then** simulate p; add p to *plan*
       else
       * create query template $q = \langle I, O, E, S \rangle$, and corresponding abstract process p_q where $I = r.inputs$, $O =$ missing inputs of p, $E =$ unsatisfied preconditions of p
       * simulate p_q in S; add q and p_q to *plan*
3. **Reconcile outputs and effects of** r**:**
   - **if** ($\forall o \in r.outputs$ available in S) and ($\forall effect \in p.effects$ satisfied in S) **then**
     - add *plan* to *plans* // *the reconciliation of r and p is finished*
   - **else**
     - *Goals*= transform missing outputs & effects to goals
       // e.g., (RequesterGoal goal) => (Goal goal)
     - **if** *solveGoals*$(Goals, S, plan, maxPlanLength)$ **then** add *plan* to *plans*
       else
       * *duplicate the plan and continue with another process in the providers model*
         · *newPlan = plan*
         · **foreach** $a \in Next(S)$ **do**
           reconcileAtomicProcess$(r, a, S, plans, newPlan, maxPlanLength)$
       * *create an unbound plan*
         · create query template $q = \langle I, O, E, S \rangle$, and corresponding abstract process p_q where $I =$ outputs produced by *plan*, $O =$ missing outputs of r, $E =$ unsatisfied effects of r
         · add q and p_q to *plan*; add *plan* to *plans*

---

degree of match between individual plan steps and required information that the step is supposed to provide. In fact, we assign a *faulty factor* to each step of the plan: the better the degree of match of the step, the lower is its faulty factor. The faulty factor of the overall plan is gained as a sum over all its steps. When it comes to the execution, the plans are considered in the descending order according to their faulty factor. This means that short plans with good quality are preferred.

5   Data Mediators Discovery

Discovering new data mediators requires two questions to be answered. First, the specific *discovery requirement* needs to be gained, and second, this requirement must be translated into *concrete queries* that will be sent to the discovery service. We use query templates defined in the previous section to capture discovery requirements. The query template is derived from the specific mismatch between the reconciled requester's request r and the process p encountered in the execution state S.

Based on the query template, concrete queries which will be sent to the discovery service need to be formulated in the *bindPlan* procedure. A straightforward idea would be to use the query template as it is. However, the vast majority of discovery services implement a matching algorithm with the assumption that an advertised service matches a request (query) when all the outputs of the request are matched by the outputs of the advertised service, and all the inputs of the advertised service are matched by the inputs of the request [7]. Thus, typically, only a single service satisfying a query can be returned while service combinations are not allowed. Such an assumption makes sense when standalone services need to be discovered. In our case, however, this assumption is too restrictive since we are not necessarily looking for one service only. On the contrary, often in the mediation scenario one specific gap identified by the process mediation algorithm can be bridged only by using a combination of several services.[2]

To deal with the problem of too restricting matching assumptions we can either generate many sub-queries out of the query template to poll the discovery service, which we want to avoid because of time constrains. Or we have to relax the matching assumptions. Benatallah et. al. in [12] propose an approach that allows a combination of several services to satisfy the service request. Their algorithm based on request rewriting guarantees that an optimal combination covering the request will be found but it is NP-hard. We have decided to go a similar direction by allowing the combination of services satisfying the request to be returned as a relevant match — we call it a *combined match*. However, we do not insist on optimality. We prefer the coverage instead of optimality, since the planning algorithm takes care about finding the optimal combination.

When answering a combined match query, the discovery service first finds a set of services that together produce the required outputs and effects (i.e., any service producing some of required outputs or effects is a good candidate). In the next step, out of these candidates, if more candidates are available producing the same outputs, those are preferred that use only inputs specified in the query. Since no real composition is involved such an approach is very efficient (assuming that appropriate index structures were precomputed during the service registration with the discovery service).

Assuming the discovery service supports a combined match, the PMA discovers new mediators in the *bindPlan* procedure in two steps as shown in Algorithm 4. It starts with an exact query request in the form of the query template (step 1.1). If some service matching the query is returned, PMA just binds it in the plan in the place of the corresponding abstract process. Otherwise, the combined match query in the same form is sent to the discovery service (step 1.2). Returned data mediators are transformed into planning operators and the *solveGoal* planning method is re-run with the state S saved in the query template. The produced plan is plugged in the place of the abstract process.

---

[2] We believe it is actually the same case in service composition in general if discovery of composed services needs to be integrated.

Algorithm 4. *bindPlan*(M), $M = \langle P, Q \rangle$ unbound reconciliation plan

---

1. **foreach** $p_q \in P$, p_q abstract process, $q = \langle I, O, E, S \rangle$ query template associated with p_q **do**
   1.1 **if** *mediators* $= askDiscoveryExact(q) \neq \emptyset$ **then**
       replace p_q in P with best matching $a \in$ *mediators*
   1.2 **elseif** *mediators* $= askDiscoveryCombined(q) \neq \emptyset$ **then**
       - transform *mediators* to planning operators and add them to *Plans Library*
       - *Goals* = transform q to goals; *newPlan* $= \emptyset$
       - *solveGoals*(*Goals*, *S*, *plan*, *maxPlanLength*)
       - replace p_q in P with *newPlan*
   1.3 **else return failed**
2. **return** $M' = \langle P, \emptyset \rangle$

---

6   Recovery

For several reasons comprehensive recovery mechanisms need to be incorporated into the PMA. First, some of many choices that the PMA makes are rather heuristic and might not work correctly during the execution. For example, the preconditions of data mediators can fail during the execution or a wrong data mediator or reconciliation plan might be selected. Also many services might be unavailable, unreliable or might just fail for other reasons during the execution. Because we do not want these failures to break the overall mediation process and also because we want to shield the requester from failures, the PMA must be able to recover from failures.

To make the process mediation robust, we introduced three levels of recovery. The lowest level deals with local and relatively inexpensive recovery of individual service calls. The next level supports recovery by trying to apply alternative reconciliation plans found by the reconciliation procedure, and finally, the highest level attempts to recover from failures by exploring possible alternative branches of the provider's process model. When a failure occurs, the recovery starts with the lowest level by trying to recover locally and only if it does not succeed, the control is passed to higher levels which are computationally more expensive and less predictable. Although every higher level introduces more global and less deterministic way of recovering from failures, it is necessary to apply these more global means of recovery to deal with possibly wrong choices of the PMA caused by missing information about the requester's process model.

The following paragraphs give more information about each level of recovery supported by the PMA. Generally, the recovery mechanisms are based on applying pre-canned fault handler templates employing recovery actions such as *replace, retry, replaceByEquivalent*, and *compensate* [8].

6.1   Individual Service Level Recovery

The PMA deals with individual service failures by automatically attaching fault handlers to individual service calls. The following example demonstrates one such a fault handler that is triggered when service call runs out of time.

```
FaultHandler(ServiceTimeoutException){  retry(2){
  FaultHandler(ExceptionEvent){ replaceBy(AlternativeService); }}}
```

The service to which this fault handler is attached is retried at most twice and if the retry fails, the original service is replaced by an explicitly specified replacement service *AlternativeService*. In this example, the *AlternativeService* was discovered as an equivalent service during the discovery and planning phase. When no alternative is known to the PMA the *replaceByEquivalent* recovery operation is used instead and a possible substitute service is discovered dynamically during the runtime.

6.2   Recovery of Reconciliation Plans - Localrecover(P, S)

When some part of the reconciliation plan fails and the recovery on the individual service level does not help, the plan as a whole needs to be recovered. In such a case the *compensate* recovery action is used to undo the the effects of the plan and possibly an alternative plan is tried. The following fault handler is a prototype associated with every reconciliation plan:

```
FaultHandler(ExceptionEvent){ compensate; }
```

The compensation of the whole plan might not be possible if it is not possible to undo the effects of each step in the plan. However, undo operations must be provided only for *world affecting* actions, i.e., actions with some effects, while the *information gathering* actions do not need to be undone. Since only the minority of web service calls are world affecting, there is a big chance that the compensation of the whole reconciliation plan will succeed even when the undo operations are not known for every action in the plan.

6.3   Global Choices Recovery - Globalrecover(S)

If none of the previous recovery levels succeeds, the PMA will try to apply the recovery on the global level by exploring possible branches in the provider's process model. There might be places in the provider's process model where the PMA has a choice among several alternative branches which are not distinguishable because of the local decision making. For example, when each branch at some choice place starts with an atomic process with exactly the same set of inputs and outputs the PMA will not be able to decide which branch to prefer. In such a case, one branch is selected by the PMA and the other possible branches are remembered together with the choice point for the possible global recovery.

The *globalRecover* procedure serves as a last resort in the recovery and its goal is to recover from a situation when a wrong branch of the provider's process model was taken. The idea is to backtrack to the nearest branching point (place with several choices) while compensating the executed actions in the execution history. If the backtracking succeeds, an alternative branch and an alternative reconciliation plan is chosen if it exists. Otherwise the backtracking continues to the next branching point.

Since the *globalRecover* can include undoing several already finished requester's actions, it needs to be constrained to avoid incorrect behavior. Specifically, if any of the actions that need to be compensated is world affecting, the requester is explicitly notified about the problem and asked whether the compensation and the backtracking can be performed. Requester will be notified also in the situation when the services executed along the alternative branch produce different results than those produced while executing the original branch.

7   Related Work

The work in [13] provides a conceptual underpinning for automatic mediation. In [14] Cimpian at. al. solve the runtime mediation between two WSMO based processes. Besides structural transformations (e.g., change of message order) also data mediators can be plugged into the mediation process, however, recovery and discovery are not addressed at all. Aberg et. al. [15] describe an agent called sButler for mediation between organizations' workflows and semantic web services. The mediation is more similar to brokering, i.e., having a query or requirement specification, the sButler tries to discover services that can satisfy it. The requester's process model is not taken into considerations. OWL-S broker [16] also assumes that the requester formulates its request as query which is used to find appropriate providers and to translate between the requester and providers. In [17] and [18] authors describe the IRS-III broker system based on the WSMO methodology. IRS-III requesters formulate their requests as goal instances and the broker mediates only with providers given their choreographies (explicit mediation services are used for mediation). Brambilla at. al. [19] apply a model-driven approach based on WebML language. Mediator is designed in the high-level modeling language which supports semi-automatic elicitation of semantic descriptions in WSMO. In [9], data transformation rules together with inference mechanisms based on inference queues are used to derive possible reshapings of message tree structures. An interesting approach to translation of data structures based on solving higher-order functional equations is presented in [10] while [20] argues for published ontology mapping to facilitate automatic translations.

8   Conclusions and Further Work

In this paper we dealt with mediation mechanisms of two OWL-S process models operating in dynamic open environments. We described algorithms based on the analysis of provider's and requester's process models for finding mappings between them and for performing the runtime mediation. The algorithms combining planning, semantic reasoning, discovery of missing data mediation services and recovery mechanisms are embedded in the Process Mediation Agent that provides the mediation services. The main advantage of our approach, besides enabling the interoperation of requesters and providers, is the capability of the PMA to operate in real conditions where failures and changes of the environment must be taken into account. Due to several levels of recovery mechanisms employing dynamic recovery and built-in heuristics, the PMA is able to recover when possible and its performance degrades gracefully when the environment changes or no simple recovery is possible.

In the paper, we focused on the mediation problem itself and we did not discuss many practical issues related to the asymmetric mediation scenario such as security, hosting of the PMA, etc. Let us discuss briefly the hosting question. In general, depending on the particular application domain, the PMA can be deployed either as part of the provider's infrastructure, requester's infrastructure or as part of the middle layer in between. In all cases, there are very strong incentives for hosting the PMA related to achieving interoperability. Hosting the PMA on the side of provider might allow new partners to interact with the provider. From the requester's perspective, hosting the PMA makes a good sense when some application needs to be extended by adding a new provider or when an existing provider needs to be replaced by a new one. In such a case the PMA can be used on the requester's side as a smart adapter to bridge the possible incompatibilities. Finally, the PMA can find its role in the infrastructure of enterprises such as mobile operators which provide access to services of third parties to their final customers.

As part of the work on runtime mediation algorithms we have identified several areas which deserve more attention. Specifically, we have proposed the *combined match* as an extension to the standard matching degrees used in discovery services. Our initial solution designed as an eager algorithm works quite well for our purpose, but it does not provide any optimality guaranties. Also this solution does not allow to discover proper compositions of services that would bridge an identified mismatch. We consider exploring alternative types of matches such as a *composed match*. Another area worth of our interest are adaptive mechanisms for improving the PMA's performance over the time. For example, right now the PMA stores all reconciliation plans that proved to work in the Plans Library which would not be a good solution if PMA lived for a long time. We plan to experiment with approaches known for example from case based learning to optimize the storage capacity and the retrieval time of plans.

References

1. Akkiraju, R., Farrell, J., Miller, J., Nagarajan, M., Schmidt, M.T., Sheth, A., Verma, K.: Web Service Semantics - WSDL-S (2005), http://www.w3.org/Submission/WSDL-S/
2. The OWL Services Coalition: Semantic Markup for Web Services (OWL-S), http://www.daml.org/services/owl-s/1.1/
3. Roman, D., Keller, U., Lausen, H., de Bruijn, J., Lara, R., Stollberg, M., Polleres, A., Feier, C., Bussler, C., Fensel, D.: Web Service Modeling Ontology. Applied Ontology 1(1), 77–106 (2005)
4. Sycara, K., Paolucci, M., Ankolekar, A., Srinivasan, N.: Automated discovery, interaction and composition of semantic web services. Journal of Web Semantics 1 (1), 27–46 (2004)
5. Vaculín, R., Sycara, K.: Towards automatic mediation of OWL-S process models. In: 2007 IEEE International Conference on Web Services, July 9-13, 2007, pp. 1032–1039. IEEE Computer Society, Los Alamitos (2007)
6. Wong, H.C., Sycara, K.P.: A taxonomy of middle-agents for the internet. In: ICMAS, pp. 465–466. IEEE Computer Society Press, Los Alamitos (2000)
7. Paolucci, M., Kawmura, T., Payne, T., Sycara, K.: Semantic matching of web services capabilities. In: First Int. Semantic Web Conf. (2002)
8. Vaculín, R., Wiesner, K., Sycara, K.: Exception handling and recovery of semantic web services. In: Fourth International Conference on Networking and Services. IEEE Computer Society Press, Los Alamitos (2008)
9. Spencer, B., Liu, S.: Inferring data transformation rules to integrate semantic web services. In: International Semantic Web Conference, pp. 456–470 (2004)
10. Burstein, M., McDermott, D., Smith, D.R., Westfold, S.J.: Derivation of glue code for agent interoperation. Autonomous Agents and Multi-Agent Systems V6(3), 265–286 (2003)

11. Paolucci, M., Ankolekar, A., Srinivasan, N., Sycara, K.P.: The DAML-S virtual machine. In: Fensel, D., Sycara, K.P., Mylopoulos, J. (eds.) ISWC 2003. LNCS, vol. 2870, pp. 290–305. Springer, Heidelberg (2003)
12. Benatallah, B., Hacid, M.S., Rey, C., Toumani, F.: Request rewriting-based web service discovery. In: Fensel, D., Sycara, K.P., Mylopoulos, J. (eds.) ISWC 2003. LNCS, vol. 2870, pp. 242–257. Springer, Heidelberg (2003)
13. Wiederhold, G., Genesereth, M.R.: The conceptual basis for mediation services. IEEE Expert 12(5), 38–47 (1997)
14. Cimpian, E., Mocan, A.: WSMX process mediation based on choreographies. In: Bussler, C.J., Haller, A. (eds.) BPM 2005. LNCS, vol. 3812, pp. 130–143. Springer, Heidelberg (2006)
15. Aberg, C., Lambrix, P., Takkinen, J., Shahmehri, N.: sButler: A Mediator between Organizations Workflows and the Semantic Web. In: World Wide Web Conference workshop on Web Service Semantics: Towards Dynamic Business Integration (2005)
16. Paolucci, M., Soudry, J., Srinivasan, N., Sycara, K.: A broker for owl-s web services. In: Cavedon, M., Martin, B. (eds.) Extending Web Services Technologies: the use of Multi-Agent Approaches. Kluwer Academic Publishers, Dordrecht (2005)
17. Cabral, L., Domingue, J., Galizia, S., Gugliotta, A., Tanasescu, V., Pedrinaci, C., Norton, B.: IRS-III: A Broker for Semantic Web Services Based Applications (2006)
18. Domingue, J., Galizia, S., Cabral, L.: Choreography in irs-iii - coping with heterogeneous interaction patterns in web services. In: Proc. 4th Intl. Semantic Web Conference (2005)
19. Brambilla, M., Celino, I., Ceri, S., Cerizza, D., Valle, E.D., Facca, F.M.: A software engineering approach to design and development of semantic web service applications. In: Cruz, I., Decker, S., Allemang, D., Preist, C., Schwabe, D., Mika, P., Uschold, M., Aroyo, L.M. (eds.) ISWC 2006. LNCS, vol. 4273, pp. 172–186. Springer, Heidelberg (2006)
20. Burstein, M.H., McDermott, D.V.: Ontology translation for interoperability among semantic web services. The AI Magazine 26(1), 71–82 (2005)

Towards an Emergent Taxonomy Approach for Adaptive Profiling

Sylvain Videau, Sylvain Lemouzy, Valérie Camps, and Pierre Glize

IRIT - Paul Sabatier University
118 route de Narbonne
31062 Toulouse cedex 09
France
{videau,lemouzy,camps,glize}@irit.fr

Abstract. The omnipresence of data processing and mobile telephony in our life (computers, PDA, GSM, GPS...) along with the evolution of wireless technologies opens the door towards new habits. To avoid being submerged by too much information it is necessary to equip each electronic component present in user's daily life with capacities to take into account his needs according to his actions, to assist him while learning and anticipating on his behavior in the most autonomous way. Personalization is clearly situated in this objective; it enables a user profile construction which has to dynamically evolve. It also has to take into account new preferences, needs and interests of this user and to forget old ones. This paper proposes a local, cooperative and real-time multi-agent approach to build adaptive and incremental profiles. First, documents are sequentially parsed, which leads to the construction of a Temporary Terminological Network (TTN). This Network is then merged with other document's extracted networks, in order to create a Permanent Terminological Network (PTN), relevant to the studied collection and used to index this collection thanks to a clustering approach. Preliminary results of the built system are then presented as well as perspectives.

1 Introduction

AgentLink Roadmap [1] asserts that agents have their place in spreading fields such as Web Services, Semantic Web, Peer-to-Peer, Grid Computing, Ambient Intelligence, Self* Systems, etc. These domains display distributed services, evolving in an open and a dynamic environment; we call them Information Systems (IS). Agents act on behalf of services by notably managing the access to their offers. They also act on behalf of users, by taking part in the localization and delivery of desired services. Agents must then have the capability to understand real user/service requirements. All these fields are faced with numerous problems, such as the difficulty of the expected information location, the volatility of the services and the quality of the retrieved information. These problems are difficult to solve because of the openness and dynamics of the environment in which they evolve: it is thus necessary to manage locally the end-users/services

R. Kowalczyk et al. (Eds.): SOCASE 2008, LNCS 5006, pp. 118–133, 2008.

representations according to the interactions evolution and the execution context. Adaptive profiling of end-users/services is an inescapable tool for any future treatment (such as dynamic user-service mapping) but is not sufficient. An adaptive response to the problems induced by the dynamics and the heterogeneity of such systems (such as workload, failures and interoperability of the components, as well as the appearance and the integration of new services...) becomes also necessary. We propose to tackle these real-time adaptation problems in an integrated way, by a local adaptive multi-criteria management of the Quality of Service (QoS) provided by these systems.

1.1 Quality of Service in Information Systems

Ideally an information system (IS) has to achieve its functional adequacy during its activity, i.e. it has to produce the function for which it was conceived from the point of view of an external observer knowing its finality. But this functional adequacy cannot always be clearly and precisely expressed, especially in complex systems, such as IS. A classical top-down approach to manage and to achieve this functional adequacy is impossible to implement; that is why new approaches need to be defined. If we consider an IS as being a system delivering a service to an end-user, its functional adequacy can be considered as the "Quality of Service" (QoS) produced by the system. Generally speaking QoS has two constituents:

1. qualitative (functional) properties, defining how well the retrieved information matches the intended information such as precision, recall and noise and
2. quantitative (non functional) properties, ensuring an effective flow in terms of end-to-end delay and including properties such as security, reliability, interoperability, bandwidth.

This QoS cannot be precisely expressed because of the multiple criteria to take into account in IS, but it has, at least, to satisfy two general aspects:

1. a system has both to fulfill the objectives a designer assigned to it and to give end-user satisfaction and
2. a system has to take into account some specific constraints such as standards (for Peer-to-Peer, Grid Computing, Web Services...), permissions (for Information Retrieval, Web Services...), CPU performances, etc.

We consider that, faced with such a diversity of criteria to take into account, the required QoS cannot be checked and managed by an external supervision. System entities have, in an autonomous and local way, to deal with environmental changes, according to what they perceive and their internal state. They have to adapt themselves to unforeseeable events (server breakdown, temporary unavailability of services for updating...), to the dynamics of available information and to the state of the physical resources. Thus, the learning phase is a never-ending process.

More precisely, from our point of view, the functional adequacy problem of IS is a QoS optimisation one. If we assume that this function can be specified (objective almost unreachable because of the extreme diversity of the criteria, the environmental inherent dynamics and the motivations of end-users), a learning algorithm universally optimal on all fields will be still remained to find. Under these conditions, Wolpert and Macready [2] proved that performances of all the optimisation methods using cost functions are equivalent, including the random. We could tackle this theoretical obstacle by coming up with a meta-heuristic able to find the relevance of a given optimisation algorithm for a given dataset. In this case, this meta-heuristic would become universally effective, which is contradictory with the "No free lunch theorems for optimisation". A way to tackle the limitations of these theorems is to find a relevant optimisation algorithm which does not need a cost function derived from the global criteria to optimise. We showed in previous works [3] that algorithms, which do not directly depend on the global function to obtain, are a way to dynamically implement systems able to self-adapt to their contexts. In the case of IS, these algorithms should utilise local emergent optimisation and thus only take into account local knowledge resulting from representation of their neighbourhood (including potentially users). The concept of agent thus becomes natural for a local emergent solving.

1.2   Adaptive Multi-Agent Systems to Tackle the Quality of Service

These constraints make Adaptive Multi-Agent System (AMAS) approach [4,3] especially suited to answer to such problems. Indeed it allows the design of complex systems that can be incompletely specified and for which an *a priori* known algorithmic solution does not exist. Adaptive Multi-Agent Systems provide an organisational approach enabling the construction of multi-agent systems which continuously and locally self adapt to the dynamic of their environment. In this approach, the designer has only to define when and how each agent composing the system has to locally decide to change its interaction links with other agents in order to achieve an organisation giving rise to the adequate global function, that is the expected global function produced by the system from the viewpoint of an external observer who knows its finality. In accordance with previous paragraph, local behaviours we propose to assign to agents do not directly depend on this intended global function. According to interactions between the multi-agent system and its environment, the organisation between agents emerges and constitutes an answer to unforeseeable events.

To reach this functional adequacy, we proved that each autonomous agent, following a cycle composed of three steps (perception/decision/action) has to keep relations as cooperative as possible with its social (other agents) or physical environment [3]. To lead to a coherent collective behaviour, whereas agents only seek to achieve individual goals, techniques of cooperation avoiding failure such as conflict, competition were developed. These failures are called "Non Cooperative Situations" (NCS) and can be assimilated to "exceptions" in traditional programming. The definition of cooperation we use is not a conventional one (simple sharing of resources, common work, etc.). Our definition is based on three local meta-rules the designer has to instantiate according to the problem to be solved:

Meta-rule 1 *(cper)*. Every signal perceived by an agent must be understood without ambiguity.

Meta-rule 2 *(cdec)*. Information coming from its perceptions has to be useful to its reasoning.

Meta-rule 3 *(cact)*. This reasoning must lead the agent to make actions which have to be useful for other agents and the environment.

The adopted approach is a proscriptive one because agents have, first of all, to anticipate, to avoid or to repair the NCSs. NCS occurs when at least one of the three previous meta-rules is not locally proved correct by an agent. Different generic NCSs can then be highlighted: incomprehension and ambiguity if *cper* is not checked, incompetence and unproductiveness if *cdec* is not obeyed and finally uselessness, competition and conflict when *cact* is not checked. This approach has great methodological implications: designing an AMAS consists in defining and assigning cooperation rules to agents. In particular, the designer, according to the current problem to solve, has *(i)* to define the nominal behaviour of an agent, then *(ii)* to deduce the NCSs the agent can be confronted with, and *(iii)* finally to define actions the agent must perform to come back to a cooperative state.

This approach was applied successfully to the resolution of various types of problems related to different fields (flood forecast [5], mechanical design [6], collective robotic [7], etc.). Obtained results encouraged us to promote the use of AMAS approach and to build a methodology ADELFE [8] for designing adaptive systems. ADELFE only concerns applications in which self-organisation makes the solution emerge from the interactions of their parts. It also gives some hints to the designer to tell him if using the AMAS theory is pertinent to build his application. If it is proved, ADELFE helps him to express the behaviour of the agents composing the system and the behaviour of the society formed by these agents.

In our case, we use it for an emergent problem solving to tackle a local emergent QoS optimisation. We propose to address the QoS of an IS as two problems to solve:

1. the Profile Management (PM) part which aims at representing centers of interest of end-users and services. This part can be seen as a qualitative aspect of the QoS of IS.
2. the Relationships Management (RM) part which allows relevant connection of end-users and services according to a need. This part can be seen as a quantitative aspect of the IS QoS.

These two parts are closely dependent. The RM part uses PM part to connect end-users and services which share common interests. Conversely, the PM part is improved by results coming from RM part; new established links between end-users and services imply an update of profiles owned by involved actors. In the remaining parts of the paper, terms such as end-users and services are used in a generic way; they have different characteristics according to the application domain. For example end-users and services mapping can be performed to associate a particular service with a request, to associate a corresponding service with a task to solve (Web Services, Grid Computing, Ambient Systems...).

This paper only focuses on the qualitative IS QoS based on the real-time PM of end-users and services. Some preliminary results to solve the quantitative QoS can be seen in [9]. After a brief state of the art on profiling in section 2, we present our global approach for building adaptive and incremental profiles. Sections 3 to 5 detail the three main parts of the profiles management with some associated results. We then conclude and give some perspectives to our work.

2   Adaptive Profiling

For the last few years, user profiling has become a topical research field [10,11]. This research trend comes from the field of information retrieval. Nowadays, the number of answers provided by search engines to a user's request remains high; locating relevant information in the list of returned documents is not an easy task and needs a considerable amount of time.

2.1   Related Works

User modeling can contribute to explore information sources, to deliver only the most relevant documents to a given user and to implement pull/push activities. Daniels [12] contrasts two classes of user model:

1. quantitative and empirical models which study the external behavior of a user by observing his interactions with the system, and
2. analytical and cognitive models which are interested in modeling the internal behavior of a user and try to identify the knowledge and the cognitive processes used.

Furthermore, as user's interests, preferences and tastes evolve over time, associated profiles should also be adapted in order to retain the desired accuracy in their exploitation. Techniques to adapt user profiles to new interests and to forget old ones are needed. Our work is typically situated in this context.

User profile acquisition can be performed in an explicit way, by collecting the information provided by a user via the system interface (selection of topics, definition of attributes, explicit judgment on the relevance of document...), or in an implicit and dynamic way, by observing his behavior when he is interacting with the system [13] (bookmarking, link selection, total time spent on a page...). Most existing systems use vectorial representations coupled with standard weighting schemes to draw up user profiles. Semantic representations are sometimes used too [14]. They display relations between the units of information characterizing the profile by proposing a hierarchy of concepts. They are generally based on ontologies [15] which confer a quite relative adaptation because they are dependent on a given domain. These approaches are sometimes coupled with techniques that take into account the evolution of the profile. Some systems employ learning algorithms originally from neural networks or genetic algorithms [16]. Most of these approaches, except [17], do not address their effectiveness to adapt to user's interests changes. More recent works try to take into account a temporal dimension (short/average/long terms interests) [18,19] or information related to the context of the user [15]. However these adaptive approaches rest on global solutions or base their reasoning on the expected result of the system, which makes them not easily applicable for simultaneously dealing with multiple criteria and managing unpredictable situations.

2.2 Adaptive Profile Management

In our work [20], real entities (users and services/content providers) are supposed to be described by a set of textual data (documents such as HTML pages for example). The objective of our approach consists in extracting from these documents a set of descriptors, i.e. a signature characterizing as well as possible their "semantic" content, and therefore, the centers of interest of the represented real entity. The originality of our work concerns the dynamic and incremental profile construction of a real entity from the real time perception of the environment (obtained from textual documents). The profile construction and exploitation consist of three phases, which are repeated for all documents characterizing the current real entity:

1. Sequential parsing of a new document giving lexical information about words constituting its sentences.
2. Generation of a temporary terminological network (TTN) from the agentification of words created from the previous parsing. In the case of a user representation, the TTN represents his short term preferences.
3. Fusion of the TTN with a permanent terminological network (PTN) and indexation of the new document. Fusion enriches the world representation according to the new document content. This PTN is dynamic and enables the profile extraction for the actual entity. Indexation creates a "signature", that is a set of descriptors characterizing the profile of the new document.

These three steps are detailed and illustrated in the next paragraphs.

3   New Document Parsing

The realized work is evaluated on a corpus of lemmatized articles of the daily French newspaper "Le Monde" (dealing with the architecture in Berlin, the drug in Netherlands and the French conscientious objectors). To observe the impact of the document size in our approach, a document composed of all the documents belonging to the studied collection was generated. To confirm realized observations made on these documents, Wikipedia articles dealing with Artificial Intelligence were also studied. These articles being long and written in different styles, the impact of these two factors on obtained results can be observed.

In order to reduce noise appearance during documents study (simultaneous presence of several conjugated verbs, singular and plural uses) lemmatization mechanisms are used. They enable to group terms which have a common root and to consider their meaning as identical. We choose TreeTagger[1] to lemmatize documents. An example of lemmatization is given in figure 1. As the corpus is in French, the English translation is given in italic and bracket.

It could be conceivable to let the system carry out this stage directly, without using another software in order to avoid the errors related to the mechanisms of lemmatization and not to depend on external resources. However, in order to focus on the profiling problem, the lemmatization stage is carried out upstream using an existing tool, thus providing us with a corpus of preprocessed documents.

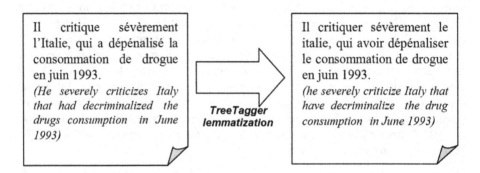

Fig. 1. Example of TreeTagger lemmatization usage

We used GraphML, an XML language to represent graphs, and Prefuse , an interactive information visualization toolkit. We chose Prefuse[2] library because of its compatibility with Java and its dynamic representation of GraphML data. We used it to import and display terminological networks generated by our system.

4   Temporary Terminological Network

The TTN and PTN are multi-agent systems composed of agentified terms parsed in documents linked by "contextual proximity" relations. These networks are constituted dynamically according to the self-organizing rules expressed in AMAS theory [4,3] and developed with the ADELFE methodology [8]. Each term making up documents is represented by a *term* agent in the system. The goal of this agent is to locally and autonomously determine its relevance in the currently studied document and to connect to other *term* agents contextually close. *Term* agents can establish two different relation types:

---

[1] http://www.ims.unistuttgart.de/projekte/corplex/TreeTagger/
DecisionTreeTagger.html

[2] Prefuse is available at the following address: http://prefuse.org

1. Spatial proximity links: these links represent the spatial neighborhood of terms, i.e. they connect *term* agents which are directly adjacent in a sentence. This type of links enables the agents to situate themselves in the studied document. Thanks to these links, *term* agents possess information enabling them to interact with their neighbors.
2. Contextual proximity links: these links point to a meaning proximity between the involved terms. These links do not only concern adjacent terms and will be used to construct the TTN and the PTN. Moreover and contrary to spatial links, these links are oriented and represent a generic/specific relation.

These two link types do not appear at the same level during the documents study, but their use is strongly connected. The terminological networks building process confirms this point.

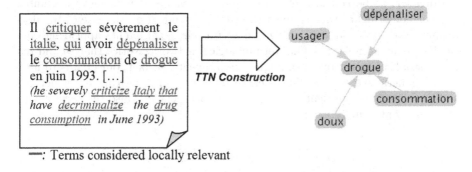

—: Terms considered locally relevant

Fig. 2. Part of TTN associated with a piece of document where terms come from

This stage takes as an input a lemmatized document and provides as an output a temporary terminological network associated with this document. The document is read sequentially. Each examined term is agentified if it is not already present in the TTN. Its goal is to find its right place in the organization; it aims to belong to a cluster composed of *term* agents which are contextually close. In order to achieve this goal:

1. A *term* agent positions itself in its environment: It observes its neighborhood to update its list of spatial neighbors.
2. A *term* agent judges of its relevance according to this new neighborhood. This estimation is performed by taking into account the number of occurrences of the actual *term* agent according to its actual neighborhood (cf. section 4.1).
3. If a *term* agent judges itself locally relevant, it tries to create contextual proximity links (cf. section 4.2).
4. When the document is entirely read, the TTN is then constructed (cf. section 4.3).

Figure 2 illustrates the four previous steps. On the left, the relevant terms found in the document containing, among other things, the text given in figure 1 are underlined. On the right, a part of the TTN focusing on the "drogue" *(drug)* term is given. "usager" *(user)* and "doux" *(soft)* terms come from other parts of the document. We will follow the highlighted words in the next figures, and in particular, their relevance change.

4.1 Local Relevance of a *Term* Agent

Let be t_a and t_b two *term* agents (t_b follows t_a in d_i), $d_i \in D$ a particular document belonging to the set of documents D contained in the collection C, $nbOcc(t_i, d_j)$ the occurrence of the *term* agent t_i in the document d_j, $r(t_i, d_j)$ the relevance of t_i in d_j, $r(t_i, PTN)$, the relevance of t_i in the actual PTN. The relevance of a *term* agent is determined by the algorithm given in figure 3.

Moreover, to be considered as relevant, a *term* agent must have a number of occurrences higher than a given threshold depending on the size of the document. This threshold was introduced to prevent *term* agents appearing once the document, to noise the final terminological network. Notice that we did not use a "stoplist" to remove meaningless *term* agents. In fact, our goal is to have a truly generic system: we thus seek to determine the *term* agents which are contextually most representative in a given document but also the *term* agents which are contextually meaningless.

```
if (nbOcc(ta, di) < 2 * nbOcc(tb, di)) then
   r(ta, di) ← true
   r(tb, di) ← false
else if (nbOcc(tb, di) < 2 * nbOcc(ta, di)) then
   r(ta, di) ← false
   r(tb, di) ← true
else if tb ∈ PTN then
   r(tb, di) ← r(tb, PTN)
else
   r(tb, di) ← true
end if
```

Fig. 3. Determination of the local relevance of a *term* agent

4.2 Contextual Proximity Links Creation

If a given *term* agent t_a judges itself locally relevant, it tries to create contextual proximity links. For that, it asks for the characteristics of its last spatial neighbor in order to determine if both appear in a similar context or not. If this neighbor t_c is not relevant, it acts in a cooperative way by providing t_a with the name of the nearest relevant *term* agent t_b it knows. A contextual proximity link is then created between t_a and t_b. This cooperative mechanism explains how contextual proximity links can connect two *term* agents which are not spatially close.

Once the two candidate *term* agents determined, their neighborhood is compared in order to create generic/specific links. Let consider t_a and t_b two *term* agents:

- $(t_a \rightarrow t_b)$ meaning that an oriented link exists from t_a towards t_b where t_a is considered as more specific than t_b
- $|nspec(t_i)|$ the number of specific neighbor *term* agents of the *term* agent t_i

Figure 4 details contextual proximity links creation.

> **if** $|nspec(t_a)| < 2 * |nspec(t_b)|$ **then**
> $(t_b \rightarrow t_a)$
> **else if** $|nspec(t_b)| < 2 * |nspec(t_a)|$ **then**
> $(t_a \rightarrow t_b)$
> **end if**

Fig. 4. Determination of contextual proximity links

We can notice that according to the context, $(t_b \rightarrow t_a)$ and $(t_a \rightarrow t_b)$ are not necessary contradictory. The presence of these two links between t_a and t_b can, for example, indicate a synonymy between these terms.

4.3 Temporary Terminological Network Construction

Obtained information during the previous steps is used to associate a confidence with relevant *term* agents and then generate the TTN. Let be:

- $nbRel(t_i)$ the number of times the *term* agent t_i has judged itself as locally relevant
- $nbOcc(t_i, d_j)$ the number of occurrences of t_i in the document d_j.

$$confidence(t_i) = \frac{nbRel(t_i)}{\sum nbOcc(t_i, d_j)}, i = 1..|D|$$

If the confidence in t_i, is greater than a given threshold, t_i is considered as a *descriptor-term* agent; this means that t_i is supposed to be representative of the document d_j content.

Once *descriptor-term* agents established, the contextual proximity links appearing during the document study enables to connect them according to relations typed generic/specific. Only the contextual proximity links existing between *term-descriptor* agents are preserved to build the TTN and are balanced with a confidence index. Let be td_a and td_b two *term-descriptor* agents:

- $(td_a \rightarrow td_b)$ meaning that an oriented link exists between td_a and td_b where td_a is a *term-descriptor* agent considered as more specific than t_b,
- $|(td_a \rightarrow td_b)|$ the number of occurrences of the oriented link between td_a and td_b,

- $|(x : x \in T, td_a \rightarrow x)|$ the total number of links where td_a is a specific boundary
- T the set of *term-descriptor* agents.

The confidence of a link is then determined by:

$$confidence(td_a \rightarrow td_b) = \frac{|(td_a \rightarrow td_b)|}{|(x : x \in T, td_a \rightarrow x)|}$$

We can observe that the previous described behaviour of *term* agents indicate that they have a quite low granularity.

Table 1. Term-descriptors confidence in the TTN of figure 2

Term-descriptor	Confidence
consommation *(consumption)*	2.00
dépénaliser *(decriminalize)*	2.00
doux *(soft)*	2.00
drogue *(drug)*	1.33
usager *(user)*	2.00

5   Permanent Terminological Network

The PTN can be seen as the aggregation of all the TTN built from documents representing a real entity. More precisely, each TTN representing a given document of this real entity will be fused sequentially with the current PTN. This PTN is also used to extract the profile of this real entity.

5.1   Fusion of TTN and PTN

Once the TTN (specific to the current document) created, it is integrated into the PTN. If the actual *term-descriptor* agent already exists in the PTN, its confidence is updated according to properties in the TTN. If it does not appear in the PTN, it is created and added to it, decreasing its confidence because its status has not already been confirmed by other documents. Then, during the fusion, each *term-descriptor* agent of a TTN tries to find its right place in the PTN in order to have cooperative relations with its environment. Three cases can be considered:

1. If a link, present in the PTN, is consolidated by a link present in the TTN, then its confidence is increased. This situation relates to links having an identical direction in the two networks.
2. If the links have a opposite direction in the two networks, we are faced to a conflict situation. This situation is solved by creating a link in the new direction met in the TTN and by decreasing the confidence of the link in the opposite direction in the PTN.

3. If the link is present in the TTN and not in the PTN, then it is added to the PTN.

The *term-descriptor* agent's confidence variations can also influence the existing links. If an agent switches from the *term-descriptor* agent status to the *term* agent status, its links are not considered as "active" because they do not relate to a relevant term, but however, they are not destroyed. Indeed, if the *term-descriptor* agent becomes later again relevant, it will always be able to position it in the PTN; the loss of history of the relations that this link has maintained during the analysis of former documents is therefore prevented.

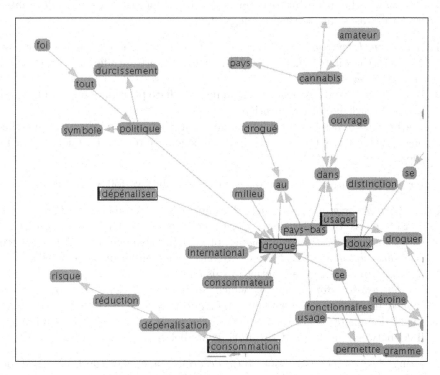

Fig. 5. PTN after fusions of 10 different documents

In Figure 5, lined in black, we find the *term-descriptor* agents that already appear in the TTN (cf. Figure 2), with an extended neighborhood, due to the study of different documents. Only *term-descriptor* agents are displayed in a PTN. Moreover, we can see in Table 2, the fusion process involved an increase in *term-descriptor* agent's confidence.

5.2 New Document Profiling

The aim of this part is to use the PTN to index documents. Indexation of a document corresponds to the extraction of its profile. We do not represent a document by a complete network extracted from the PTN, but by a set of *term-descriptor* agents delimiting areas (frontiers *term-descriptor* agents) of the terminological network, also called "clusters", which are representative of the document. Thus, only the PTN is preserved, and the superposition of the frontiers *term-descriptor* agents of a document on this network makes it possible to find the topics tackled. A cluster is a set of *term-descriptor* agents, defined by its frontiers. These frontiers are composed of the most generic and the most specific *term-descriptor* agents of the cluster. The interest of the approach by clusters is that it makes it possible to find the wholeness of the intermediary *term-descriptor* agents by a simple message passing. For this, a message is tagged "specific" or "generic" depending on whether it is sent to a more specific or a more generic *term-descriptor* agent. Three cases can be then distinguished:

1. If a *term-descriptor* agent receives only one type of message (generic or specific), then it belongs to the frontier (as a more specific or generic *term-descriptor* agent).
2. If it receives two messages, it is in an intermediate position, and can be found by the knowledge of the frontier.
3. If it receives a message which it already sent, then, a loop is located, and this one, considered as being made up of intermediate terms, does not appear in the frontier.

Moreover, the frontier construction has to check the two following rules:

R1. A *term-descriptor* agent which does not belong to the current document is considered as a frontier element of this cluster if it is recognized frontier by at least two *term-descriptor* agents present in this document. Among these *term-descriptor* agents, at least one must be a frontier of this cluster and at least one must not be a frontier of this cluster.

R2. A *term-descriptor* agent td_a which does not belong to the document is judged frontier of the cluster if $confidence(td_a \rightarrow td_b)$, with td_b belonging to the document, is at least as high as the maximum confidence found in the links between any *term-descriptor* agents belonging to the cluster.

During clusters reconstruction, *term-descriptor* agents of the PTN which do not belong to the actual document can be included in the clusters. In order to avoid a too great divergence of context and to make sure that the frontier delimits a relevant cluster, we introduced the concept of distance between the frontiers.

Table 2. Term-descriptors agents confidence in the PTN of figure 5

Term-descriptor	Confidence
consommation *(consumption)*	4.86
dépénaliser *(decriminalize)*	4.00
doux *(soft)*	2.92
drogue *(drug)*	10.98
usager *(user)*	3.00

This distance is added to make sure that the cluster is not too big and that *term-descriptor* agents composing it deal with a same context or close domains. Given a *term-descriptor* agent td_a, Fs the set of specific *term-descriptor* agents composing the frontier of the cluster and Fg the set of generic *term-descriptor* agents composing the frontier of the cluster, the distance at the frontier D is the shortest path connecting an element of Fs and Fg, passing by td_a. We can then determine a threshold to ensure that a found *term-descriptor* agent is coherent with the context defined by the cluster.

6   Conclusion and Perspectives

This paper presents a local and incremental approach to deal with the construction of adaptive profiles of real entities characterized by textual documents (such as HTML pages). The proposed multi-agent approach is based on the implementation of two different terminological networks and is totally new:

1. the first one represents the short term profile of a real user/service and is based on one document composing this user/service's competence and
2. the second represents the global profile of a real entity, based on all the documents composing this user/service's competence.

Every term in these terminological networks is an adaptive agent which tries to locally find its right place in the organization of the actual terminological network. Contrary to tf*idf which works on a closed set of documents, our approach enables to build and update, in real time, the profile of a real entity. Furthermore it is especially suited to open and dynamic environments.

Profiles have to be constructed while guaranteeing privacy of personal data. Our approach and its management is in agreement with this constraint because of its strictly local processing. Moreover, our solution could be experimented and deployed as a browser plug-in which can be used to create local and adaptive profiles improving user browsing.

The obtained encouraging preliminary results convinced us to study thoroughly the use of the AMAS for the qualitative aspect of QoS in IS. But several tasks still remain to be realized: the first one is to complete an efficient implementation of the whole proposed profiling process, notably with the indexation of a document by using frontiers of clusters.

Furthermore, to follow as faithfully as possible the AMAS approach we have to better understand what a nominal behavior means for a *term* agent or a *term-descriptor* agent. As we are situated at a very low level of cognition, such a behavior is difficult to define for the agents. That is why we are now investigating on improving the AMAS approach in order to build a system able to dynamically learn by itself rules enabling the creation of a relevant TTN. In that case the system should be able to guess predicates that rule the network creation (*term* agent creation, links creation between relevant *term* agents). With such a learning process, a *term* agent should be able to autonomously find its optimal place in the organization in order to optimize the construction and the incremental management of profiles.

References

1. Luck, M., McBurney, P., Shehory, O., Willmott, S.: Agent Technology: Computing as Interaction (A Roadmap for Agent Based Computing). AgentLink (2005)
2. Wolpert, D.H., Macready, W.G.: No free lunch theorems for optimization. Evolutionary Computation, IEEE Transactions (1), 67–82
3. Gleizes, M.-P., Camps, V., Glize, P.: A Theory of Emergent Computation Based on Cooperative Self-Oganization for Adaptive Artificial Systems. In: 4th European Congress of Systems Science (1999)
4. Gleizes, M.-P., Camps, V., George, J.-P., Capera, D.: Engineering Systems which Generate Emergent Functionalities. In: Engineering Environment-Mediated Multiagent Systems - Satellite Conference held at The European Conference on Complex Systems, Dresden, Germany, Katholieke Universiteit Leuven (2007)
5. George, J.P., Gleizes, M.P., Glize, P., Régis, C.: Real-time Simulation for Flood Forecast: an Adaptive Multi-Agent System STAFF. In: AISB 2003 symposium on Adaptive Agents and Multi-Agent Systems, University of Wales, Aberystwyth, 07/04/03-11/04/03, Society for the Study of Artificial Intelligence and the Simulation of Behaviour, pp. 109–114 (2003)
6. Capera, D., Gleizes, M.-P., Glize, P.: Self-Organizing Agents for Mechanical Synthesis . In: Di Marzo Serugendo, G., Karageorgos, A., Rana, O.F., Zambonelli, F. (eds.) ESOA 2003. LNCS (LNAI), vol. 2977, pp. 169–185. Springer, Heidelberg (2003)
7. Picard, G.: Agent Model Instantiation to Collective Robotics in ADELFE . In: Gleizes, M.-P., Omicini, A., Zambonelli, F. (eds.) ESAW 2004. LNCS (LNAI), vol. 3451, pp. 209–221. Springer, Heidelberg (2005)
8. Bernon, C., Camps, V., Gleizes, M.P., Picard, G.: Engineering Adaptive Multi-Agent Systems: The ADELFE Methodology, pp. 172–202. dea Group Pub (2005)
9. Camps, V., Glize, P.: Towards a Self-Adaptive Multi-Agent Approach for Enhancing the Quality of Service provided by Open Information Systems. In: 3rd International Conference on WEB Information Systems and Technologies (WEBIST 2007), Web Interfaces and Applications, Barcelona, pp. 295–301. INSTICC Press (2007)
10. Montaner, M., López, B., De La Rosa, J.L.: A Taxonomy of Recommender Agents on the Internet. Artif. Intell. Rev. 19(4), 285–330 (2003)
11. Brusilovsky, P., Kobsa, A., Nejdl, W. (eds.): Adaptive Web 2007. LNCS, vol. 4321. Springer, Heidelberg (2007)
12. Daniels, P.J.: Cognitive models in information retrieval -an evaluative review. J. Doc. 42(4), 272–304 (1986)
13. Lieberman, H.: Letizia: An Agent That Assists Web Browsing. In: Mellish, C.S. (ed.) Proceedings of the 14th International Joint Conference on Artificial Intelligence (IJCAI 1995), Montreal, Quebec, Canada, pp. 924–929. Morgan Kaufmann, San Francisco (1995)
14. Sieg, A., Mobasher, B., Burke, R.: Inferring users information context: Integrating user profiles and concept hierarchies. In: 2004 Meeting of the International Federation of Classification Societies, Chicago, IFCS (2004)
15. Baziz, M., Boughanem, M., Aussenac-Gilles, N.: Semantic Networks for a Conceptual Indexing of Documents in IR. In: ISPS 2005, Seventh International Symposium on Programming and Systems, Algiers, Algeria, pp. 213–224 (2005)
16. Menczer, F.: ARACHNID: Adaptive Retrieval Agents Choosing Heuristic Neighborhoods for Information Discovery. In: 14th International Conference on Machine Learning (1997)

17. Moukas, A.: User Modeling in a MultiAgent Evolving System. In: Workshop on Machine Learning for User Modeling, 6th International Conference on User Modeling, Chia Laguna, Sardinia. (1997)
18. Kilfoil, M., Ghorbani, A.: SWAMI: Searching the Web Using Agents with Mobility and Intelligence. In: Kégl, B., Lapalme, G. (eds.) Canadian conference on AI, Victoria, Canada, pp. 91–102. Springer, Heidelberg (2005)
19. Bottraud, J.C.: Un assistant adaptatif pour la recherche d'information: AIRA (Adaptative Information Retrieval Assistant). PhD thesis, Université Joseph Fourier (2004)
20. Videau, S.: Étude de la dynamique des profils adaptatifs dans un système d'informations. Master's thesis, UPS Toulouse 3 (2007)

Commitment-Based Service Coordination

Stefan J. Witwicki and Edmund H. Durfee

Computer Science and Engineering
University of Michigan
Ann Arbor, MI 48104, USA
{witwicki,durfee}@umich.edu

Abstract. We present a methodology for the composition of rich services that exhibit temporal uncertainty and complex task dependencies. Our multi-agent approach incorporates temporal and stochastic planning paradigms and commitment-based negotiation to achieve coordinated provision of services with stochastic outcomes. This is all captured within a service-choreography protocol, by which agents can request services and receive probabilistic temporal service promises, to iteratively converge on coordinated behavior. We argue that such an approach partially decouples the problems of negotiating service interactions and computing service policies, so as to more efficiently converge on good solutions.

1 Introduction

Timing can be important within and across the provision of services. For example, the outcome of a service might be needed by some deadline, or there might be a need to time the provision of complementary services to ensure that they are providing contemporaneous results. The process by which a service request is accomplished could involve stochastic latencies or uncertainty over which sequences of specific tasks will be invoked to achieve the service in various circumstances.

In this paper, we focus on the problem of coordinating services in domains involving temporal constraints and duration uncertainty. Our approach represents *commitments* between service-requesting and service-providing agents with explicit temporal and probabilistic parameters, such that at the level of coordination agents only reason about service outcomes. Agents then can use these commitments internally to guide the construction of policies about how to achieve (in the case of providers) or utilize (for requesters) these service outcomes. Our hypothesis is that effective service coordination can be achieved more efficiently by using commitments to largely decouple the service coordination and service achievement subproblems, rather than by coordinating detailed policies.

1.1 Example

Consider the simple service-oriented agent problem as viewed by a service-providing agent, as depicted in Figure 1. The service-providing agent (Agent 1)

R. Kowalczyk et al. (Eds.): SOCASE 2008, LNCS 5006, pp. 134–148, 2008.

has various (temporally uncertain) tasks that it can perform to fulfill (temporally conditioned) service requests of other agents. It provides 3 services {A, B, and C}, where A requires the completion of Task A, B requires the completion of task B, and C requires the completion of task B followed by the completion of task C. Requests come in the form of: *complete Service X by time t with probability p.* The service-providing agent can only perform one task at a time and there may be several service-requesting agents. Furthermore, we assume that execution must occur within a finite problem window of $[0, T]$ (where T=8 for the problem in Figure 1).

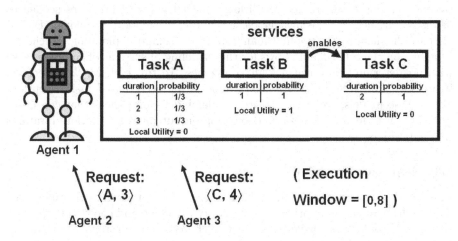

Fig. 1. Service-Oriented Coordination Example

There are two service requests in our example: one from Agent 2 and one from Agent 3. In this paper, we assume that these requests arrive sequentially, and that one request is handled before the next is considered. In coordinating its behavior with its peers, the service-providing agent needs to build a policy that coordinates the execution of its tasks so as to fulfill the incoming service requests. Several factors complicate this planning problem. There is uncertainty in the duration of task A, which may take with equal probability 1, 2, or 3 time units. Also, the successful completion of task C requires that task B be executed prior to task C. In order to plan effectively, agents must model these kinds of uncertainties and dependencies.

1.2   Related Work

The example above is one of automated service composition, a field that has been extremely active in the last several years [1,2]. Much attention has been devoted to the decomposition of complex problem solutions into the execution of individual services: from the identification of complex service combinations and orderings to the enlistment of services and management of service activities.

Two main paradigms have emerged [3]. *Service Orchestration* employs a central coordinator to invoke and combine services. *Service Choreography* instead brings services together through peer-to-peer interaction without the need for central control. Much of the recent work focuses on the problem of identifying combinations and sequences of services to be composed, but the individual services are modeled as simple processes that can be invoked when needed. There is less of an emphasis on planning the temporal interactions between more complex services involving stochastic, interdependent processes.

To solve these richer service problems, we draw upon temporal planning [4] and stochastic planning [5]. Temporal planning takes representations of task durations, temporal constraints, conditions, and effects, and produces sequences of tasks that achieve specified goals. Disjunctive Temporal Problems (DTPs) [6,7] are the most recent popular manifestation of temporal planning. Stochastic Panning, on the other hand allows for the modeling of systems of complex tasks with nondeterministic durations and outcomes. To this end, Markov Decision Processes (MDPs)[8] provide powerful models for agent-based task execution. Using these models, agents can represent flexible action polices that can interleave activities for different tasks depending on outcomes and action durations that actually occur.

1.3   Approach Overview

In order to compose services and achieve their goals, our service-oriented agents require the completion of other agents' services. To plan and coordinate the execution of these services, we utilize a service choreography protocol. As shown in Figure 2, service-requesting agents submit requests to service-providing agents. The requests are dealt with through negotiations between the requester and provider that potentially end in agreements for service provision.

As we describe in the sections that follow, for steps 1 and 4, service-requesting agents employ temporal and stochastic planning to reason about the timing of when the services are needed in order for their own temporally constrained

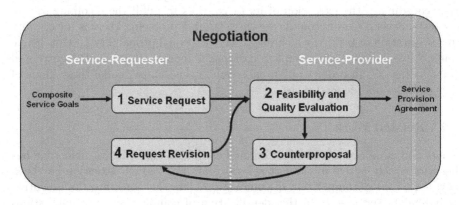

Fig. 2. Negotiation Protocol

goals to be met. Because of the stochasticity and service dependencies, service-providing agents also employ temporal and stochastic planning techniques in steps 2 and 3 to decide what services can be provided at what times and with what likelihoods. We assume that the service agents are fully cooperative, such that agents will perform tasks to achieve mission objectives and maximize their collective utility. For simplicity, we consider the agents' collective utility to be the sum of their individual rewards.

We begin by presenting a MDP modeling framework to represent service agents' tasks, execution, and commitments for service provision. Next, we provide a methodology for service-provider reasoning: how to constrain its policy-formulation based on its commitments and in doing so evaluate the feasibility of commitments (step 2), and how to search the space of commitment values when formulating counterproposals (step 3). We present a corresponding methodology for service requesters to evaluate counterproposals and formulate new service requests (steps 1 and 4). Having brought together all of the steps of the negotiation protocol, we discuss how the overarching problem of coordinating services activities of the system of agents may be achieved through commitment convergence. Finally we present our ongoing efforts to analytically and empirically evaluate this approach.

2   Agent Models

Here we present the details of our framework for modeling service agents and their individual tasks that must be executed to provide their services.

2.1   Markov Decision Processes

Because of the uncertainty, conditional constraints, and temporal constraints of the service-providing problem from Figure 1, we model the behavior of a service agent using a Markov Decision Process (MDP). In review, a classical MDP can be described by a 4-tuple $\langle S, A, P, R \rangle$, where: S is a finite set of world states, A is a finite set of actions, P is the transition probability function $P : S \times A \times S \rightarrow [0, 1]$, and R is the reward function $R : S \times A \rightarrow \mathbb{R}$.

The solution of a MDP is a policy π, which may be described as a mapping of states to probability distributions over actions ($\pi : S \times A \rightarrow [0, 1]$). An optimal policy π^* maximizes the agent's total expected reward. There are several common approaches for computing the optimal policy π^* of a MDP [9]. These include Dynamic Programming (i.e. policy iteration, value iteration), Monte Carlo methods, and reinforcement learning. In this paper, we use the Linear Programming (LP) approach [10,8]. A MDP as described above can be formulated as a Linear Program:

$$
\max \sum_{i,a} x_{ia} R(i,a) \left| \begin{array}{l} \forall j, \sum_{a} x_{ja} - \sum_{i,a} x_{ia} P(j|i,a) = \alpha_j \\ \forall i \forall a, x_{ia} \geq 0 \end{array} \right. \tag{1}
$$

where α_j denotes the probability of starting in state j, and the x_{ia} variables, often called *occupancy measures*, denote the total expected discounted number of times action a is performed in state i. The optimal policy π^* can be expressed simply in terms of the optimal occupancy measures. So upon solving this LP, $\pi^*_{i,a}$ is easily computed along with the expected utility $EU(\pi^*)$.

2.2 Modeling Tasks with Temporal Constraints and Uncertainty

We represent a service-providing agent's problem with a MDP. The states in the state space are modeled in terms of the features relevant to the services that the agent provides. Since service requests come with a time constraint, an agent should know, in any given state, how much time has passed. So one feature of a MDP state is its *time*. Further, to model probabilities of service completion in subsequent states, and to preserve the Markov property, we incorporate into a state the *task status*: for each task, whether the task has completed successfully, whether it has not yet been attempted, or whether it is in the midst of execution and if so the time at which it was started. The actions available to the agent are to start tasks ($START$-$task$-x) or to do nothing ($NOOP$). For each task, a reward equal to the local utility of that task is assigned to any state in which the task has just been completed.

It is now straightforward to construct a MDP model that corresponds to an agent's task execution and service provision. Figure 3 shows part of the MDP for the service-providing agent in our example problem. At time step 0, the agent can either execute Task A, Task B, or NOOP. It cannot execute Task C because C requires that B be previously completed. Upon executing B, the agent immediately transitions into a state in which B is "(F)inished". But upon executing A, the agent nondeterministically finishes A with probability 1/3 and transitions accordingly. The states at times 0, 1 and 2 are shown.

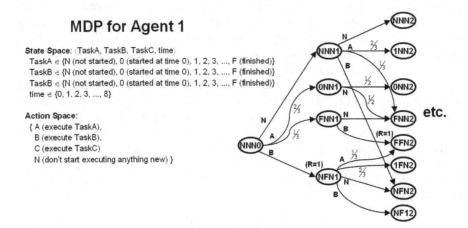

Fig. 3. Example MDP for Service-Providing Agent

2.3   Commitments

We extend previous work on commitment-based MDP coordination [11] to include a temporal component that is particularly relevant for temporally uncertain services. Our commitment-based coordination methodology is centered on requests and promises. The service-requester requests that services be provided, and the service-provider promises to deliver services to the service-requester. Turning again to our example problem, the first request (from Agent 2) is for Service A to be completed by Time 3. This request can be fulfilled if the service provider (Agent 1) executes Task A at time 0. Task A may finish earlier than time 3, but its maximum duration is 3, so it can't finish any later. Thus Agent 1, if willing, can make a promise to Agent 2 to complete A (with certainty) no later than time 3. We call this promise a commitment because, if it is accepted by Agent 2, Agent 1 is required execute a policy that will deliver Service A with certainty no later than time 3.

Next, consider the second request (from Agent 3) for Service C to be completed by time 4. If Agent 1 is committed to completing A by time 3, it cannot promise to complete C by time 4. Task C deterministically takes 2 time units to complete and requires that B be completed prior to it starting. If Agent 1 begins tasks B and C as soon as it can after completing A, in the worst case, C will not complete until time 6. There is, however, a $\frac{1}{3}$ probability that Task C will be completed by time step 4 (when Task A takes 1 time step). This gives rise to the notion of a probabilistic temporal commitment.

Definition 1. *A **probabilistic temporal service commitment** $C_{ij}(s) = \{\rho, t\}$ is a guarantee that agent i will perform (for agent j) the actions necessary to deliver service s by time t with probability no less than ρ.*

These probabilistic commitments allow agents to makes promises to each other even in the event that they cannot fully guarantee service provision. It can be extremely beneficial to model the inherent service uncertainty in this way. As we will discuss later on, Agent 3 may find it acceptable to receive Service C with a lower probability if it will be received at a desirable time.

3   Service-Provider Commitment Reasoning

Now that we have the models for service agents, tasks, and commitments, we can describe the inner workings of the negotiation protocol introduced in Figure 2. We begin by showing how service-providing agents can evaluate the feasibility of a received request (step 2 of the protocol). We then discuss how alternative commitments can be proposed (step 3).

3.1   Forming Commitment-Constrained Policies

To adhere to its probabilistic temporal commitments, a service-provider needs to calculate a policy that keeps its promises. Prior approaches for doing this have introduced extra rewards for reaching commitment-satisfying states into the MDP, and then solved the MDP in standard ways [12]. Our previous work on (non-temporal) probabilistic commitments developed a more effective alternative by constraining the space of policies rather than doctoring the rewards (to bias policy transitions) [11]. We extend our previous work to find policies constrained by *temporal* probabilistic commitments.

Our solution uses the linear programming approach described in Section 2.1. We directly modify the MDP linear program from Equation 1 to constrain the solution policy to adhere to a set of *temporal* probabilistic commitments:

$$
\max \sum_i \sum_a x_{ia} R\left(i,a\right) \left| \begin{array}{l} \forall j, \sum_a x_{ja} - \gamma \sum_{a,i} x_{ia} P\left(j|i,a\right) = \alpha_j \\ \forall i \forall a, x_{ia} \geq 0 \\ \forall s \qquad \sum_{\{i|\{time(i)=t_s \wedge Status_s(i)=F\}\}} \sum_a x_{ia} \geq \rho_s \end{array} \right. \tag{2}
$$

Equation 2 includes a third constraint, requiring that the committing agent's policy visit states with $time = t_s$ and a $Finished$ status of service s with probability no less than ρ_s.[1] Solving the new linear program will yield a policy that is optimal for the committing agent with respect to its commitments to other agents if such a policy exists. If no such policy exists, the agent is overcommitted, and so the Linear Program is overconstrained and has no solution. In this case, the LP solver will output that there is "NO SOLUTION".

3.2   Pruning Commitment Times

When a service request cannot be honored as requested, the LP formulation will find no solution. Rather than replying "no" to the requester, the protocol expects the provider to supply one or more counterproposals that represent alternative requests that it could commit to fulfilling (step 3 in Figure 2). In considering the space of possible counterproposals, not all commitment probabilities and times need be considered. In the following sections, we present some techniques to prune suboptimal values from the space of potential commitment values. First we focus on the time dimension of the space, and later on the probability dimension.

Recall that, for the service-providing agent, a commitment pertains to the completion of one of its tasks. Each task has a certain discrete probability distribution over durations. So, to pick a time to promise the task completion with any probability greater than zero, it does not make sense to consider times that are less than the smallest positive probability duration.

In the example problem, our agent cannot complete Task A before time step 1. For tasks that depend on other tasks, we can push the earliest commitment time further forward by adding the minimum durations of all dependent tasks. Task C depends on the completion of Task B, so the earliest time that should be considered for completing C is $2 + 1 = 3$.

---

[1] Because states are time indexed, no state is visited multiple times in any execution.

More sophisticated temporal reasoning can push the earliest commitment time back even further. Given an existing commitment by Agent 1 to deliver Service A at time 3, we can deduce that Task A must be started at time 0 and cannot finish any earlier than time step 1. So given previously established commitments, C should not be committed to any earlier than time 4.

3.3   Bounding Commitment Probabilities

Having reduced the commitment space with respect to the time dimension, let us now consider the probability dimension. If the service-providing agent makes a commitment to completing Task A at time 2, it makes sense to set the commitment probability equal to the probability with which it can complete A in two time steps or less: 2/3. If the agent promises a higher probability, it will not be able to meet its commitment. We say that 2/3 is the maximum feasible probability for Agent 1's commitment to providing A at time 2.

Definition 2. *The **maximum feasible probability** of a commitment C made at time t is the highest commitment probability than can be achieved by time t by any policy of the supporting agent.*

The maximum feasible probability of commitment to service s_k can be computed using a linear program, slightly modified from Equation 2, that takes as input the service-providing agent's local MDP with all previously made commitments set to their promised values and (using occupancy measures) maximizes the probability of s_k being delivered at the given time:

$$\max \rho_{s_k} \quad \left| \begin{array}{l} \forall j, \sum_a x_{ja} - \gamma \sum_{a,i} x_{ia} P\left(j|i,a\right) = \alpha_j \\ \forall i \forall a, x_{ia} \geq 0 \\ \forall s \displaystyle\sum_{\{i|\{time(i)=t_s \wedge Status_s(i)=F\}\}} \sum_a x_{ia} \geq \rho_s \end{array} \right. \tag{3}$$

In this new linear program, ρ_{s_k} is a probability variable (unlike the rest of the $\{\rho_s\}$ constants) and the solution maximizes that probability instead of maximizing local utility as we did in Equation 2.

When a request is deemed infeasible, the service-provider can, in this fashion, calculate a *maximum feasible probability* boundary across all relevant commitment times. Consider our example problem shown in Figure 1. The first request for A to be completed by time step 3 can be honored and a commitment ($C_{12}(A) = \{\rho = 1.0, t = 3\}$) formed. But next, the service provider receives a request from Agent 3 to deliver C by time step 4 (with implicit probability 1). Given the first commitment made to Agent 2, a commitment $C_{13}(C) = \{\rho = 1.0, t = 4\}$ is not feasible. This is shown in Figure 4.

The service-provider could, in principle, calculate the entire *maximum feasible probability* boundary. But in counterproposing, it makes sense to use the time

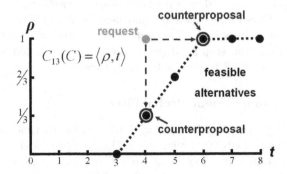

Fig. 4. An Example of Commitment Counterproposal

and probability of the request as a basis for feedback. As shown in Figure 4, C can be delivered by a later time, 6, with the same probability as the original request, yielding alternative commitment $C'_{13}(C) = \{\rho = 1.0, t = 6\}$. Or C can be delivered at the same time as the request but with smaller probability, yielding $C''_{13}(C) = \{\rho = \frac{1}{3}, t = 4\}$. These two counterproposals give the requester some idea of the boundary capabilities of the provider. Other points along the boundary could be provided, depending on the details of the negotiation protocol.

4   Service-Requester Commitment Reasoning

4.1   Commitment Consideration and Request Formulation

We now discuss how a service-requesting agent like Agent 3 would process the commitments counterproposed by the service-provider in its negotiations (step 4 in Figure 2). We assume, in this discussion, that the service requester has a way of evaluating the utility of a particular commitment. For example, if the requester models its own behavior with a MDP policy, a commitment can be captured by transitions, with the ρ values corresponding to transition probabilities and the t values dictating the states affected by the transitions [11]. The utilities of different commitments can then be computed by solving the MDPs and calculating the expected utilities of their respective solution policies.

One very simple method of formulating a new request is to evaluate each counterproposal, identify the best one, and request it. In our example, Agent 3 either could choose time 6 with probability 1, or time 4 with probability $\frac{1}{3}$. A slightly more advanced method for formulating a new request, though, would be to choose a commitment time and probability between the bounds of the counterproposals. The requester can very simply interpolate optimistically, computing a request, for example, for commitment $C'''_{13}(C) = \{\rho = 1, t = 5\}$. Although an interpolated commitment request is not necessarily feasible, in the worst case the provider will respond with more counterproposals. We can prove that, by iterating back and forth in this way, the potential commitment time window will narrow monotonically and (since time is discrete) the process must terminate when the requester is unable to interpolate further.

From the perspective of the service-requester, another response to counterproposals from potential service providers could be to consider them collectively, and accept multiple such proposals. In our running example, had there been a second potential provider for service C, the service request could have gone to it as well as to Agent 1. Let's say that having service C at time 4 is important for the requester. The counterproposal from Agent 1 specifies that, at time 4, there is a probability of $\frac{1}{3}$ that service C will be accomplished. If the other provider responded that, at time 4, it could provide C with a probability of $\frac{1}{2}$, then the requester has options. It could certainly choose to enlist the other agent to provide C, because of the higher probability. But, assuming that the possible providers are otherwise idle, and that they can pursue C concurrently and independently, the requester could accept *both* counterproposals, so as to increase the probability that at least one provision of C will succeed to $\frac{2}{3}$.

4.2   Request Initialization

To begin the negotiation process, the service-requester must formulate initial requests to send to the service-providers (step 1 in Figure 2). We present one method by which all requests may be initialized. A service-requester wants to formulate its best possible policy, which it can optimistically formulate by assuming that all of its commitment requests will be satisfied fully as early as it wants. That is, it can imagine that all providers will agree to commitments at time zero with probability 1, and formulate its own optimal policy correspondingly, yielding is maximal local expected utility u^*. Then, given that it knows this maximal local expected utility, the requester can then turn the optimization problem around to find the *latest* time for the commitments that can achieve this utility. We have developed a mixed-integer linear program, shown in Equation 4, for computing a policy that performs commitment-enabled actions as late as possible while maintaining that the local utility is no worse than u^*.

In Equation 4, we introduce integer variables $y_t \in 0, 1$ that can only take a value of 1 if a commitment-utilizing action is performed at or before time t with probability greater than 0. In minimizing the sum of the y values, we force commitment-utilizing actions to be performed as late as possible. Upon solving the MILP, the earliest time of such an action may be calculated by finding the first y variable that has value 1: $min_t \{y_t > 0\}$. This earliest commitment-utilization time returned by the linear program is then used as a relaxation time for the requested commitment. These relaxed requests may still be overly optimistic, but at least they do not impose unnecessarily demanding requirements on the service-providers.

$$\min \sum_t y_t \left| \begin{array}{l} \forall j, \sum_a x_{ja} - \gamma \sum_{a,i} x_{ia} P\left(j|i,a\right) = \alpha_j \\ \forall i \forall a, x_{ia} \geq 0 \\ \forall s \quad \sum_{\{i|\{time(i)=t_s \wedge Status_s(i)=F\}\}} \sum_a x_{ia} \geq \rho_s \\ \sum_i \sum_a x_{ia} R\left(i,a\right) \geq u^* \\ \forall t < T, -1 \leq \left(\sum_{\{i|\{time(i)\leq t \wedge C \in preconditions(a)\}\}} x_{ia} \right) - y_t - \varepsilon \leq 0 \\ \forall t, y_t \in \{0,1\} \end{array} \right. \tag{4}$$

5 Commitment Convergence Using the Negotiation Protocol

Each request made to a service provider may be handled using the negotiation protocol introduced in Figure 2. We have described in the previous sections methodologies for each step of negotiation that will iterate through sets of potential commitment values and eventually converge on a single agreed commitment for each requested service. As in our example problem, the service-providing agent is given a sequence of these incoming requests prior to execution. The idea is to consider each request as it comes in, and to form an agreement with the service-providing agent(s) through negotiation. Our agents therefore search the space of commitments of all service requests greedily by setting the commitments one at a time. This strategy enables much quicker commitment convergence than would an exhaustive search, but just as with other greedy methods, there is no guarantee that optimal sets of commitments won't be overlooked.

We have developed a preliminary implementation of this convergence algorithm in Java with JNI calls to cplex to solve the commitment-constrained MDPs. Figure 5 shows the runtime on a version of the example problem that is scaled up by simply stretching out the timing of all tasks[2] and extending the time horizon accordingly (from T=8 to T=96). This leads to larger MDPs, more LP constraints, and potentially more iterations of commitment requesting and counterproposing. As shown, the algorithm remains tractable for time horizons up to T=96 (at which point cplex is solving constrained MDPs with over 10,000 states), converging on commitments in a minute or less. We compare this runtime with solving a Multiagent MDP construction (in which joint actions are modeled). The Multiagent MDP scales much worse with the problem time horizon, taking hours to return a solution that the commitment-based algorithm returned in under a minute.

---

[2] Tasks maintain the same number of discrete durations, but each possible duration is scaled.

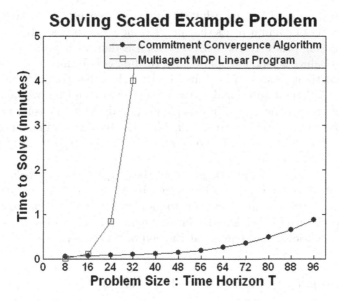

Fig. 5. Preliminary Empirical Results

These results provide very preliminary evidence that supports our hypothesis that commitment-based techniques can be more computationally efficient because they help decouple decisions about how to coordinate service provision from decisions about how to build policies that actually provide the services in stochastic domains. However, the computational benefits can come at a price in terms of the quality of the agents' joint solution. By focusing on finding a single probabilistic temporal service commitment for each request, the agents' joint policies forgo some potentially valuable flexibility. There is nothing in the commitment-based approach that precludes making multiple (conditional) commitments at different times for each request, but this would further enlarge the coordination search space, and so should be done with care.

As mentioned above, the protocol used in this paper takes a greedy approach, which can also sacrifice solution quality. When the service requests are handled in the order that they are shown in Figure 1, negotiation yields a commitments $C_{12}(A) = \{p = 1.0, t = 3\}$ and $C_{13}(C) = \{p = \frac{1}{3}, t = 4\}$. Given that completion of Task A by time 3 is worth a local utility gain of u_2 to Agent 2 and the completion of Task C by time 4 is worth a local utility gain of u_3 to Agent 3, these two commitments together provide the requesters a total expected gain of $u_2 + \frac{1}{3}u_3$. If we were to reverse the order in which the requests are considered in the example problem, the negotiation protocol brings us to a different set of commitments. A commitment $C_{13}(C) = \{p = 1.0, t = 4\}$ will be made to Agent 3 promising the completion of Task C by time step 4. But when the provider next negotiates with Agent 2, it can only make commitments involving the execution of Task A after Tasks B and C. Otherwise its first commitment would be violated. In the case of our example problem, completion of task A after time 3 does not benefit Agent 2 at all. Thus, by using this alternate request order, negotiations converge on a set of commitments that provide the requesters a total gain of u_2.

Which ordering produces the better solution is dependent upon the relative utility benefit values u_2 and u_3. Specifically, the first commitments are preferable when u_2 is worth at least $\frac{2}{3}$ of u_3, but otherwise the other commitments would be preferred. Although additional ordering heuristics could be overlaid on top of the greedy protocol described in this paper, it is difficult to ensure in general that the right ordering will be attempted. The MMDP formulation, on the other hand, always makes the correct choice and always converges on the globally optimal joint policy for the agents.

In summary, our early results suggest that the computational benefits that we seek can be achieved by using a commitment-based approach to service coordination. Further work remains, however, to more systematically characterize the potential computation benefits and solution-quality costs of the approach under a variety of conditions, as discussed at the end of this paper.

6   Factoring in Service-Provider Utility

The discussion so far has assumed that service-providers only have utility in terms of servicing requests. When service-providers also generate utility locally, the space of commitments to consider grows, because the "best" commitment in terms of maximizing *total* utility might not be along the maximal feasible probability boundary. That is, by reducing the probability with which it will satisfy a request to a less-than-maximal value, the service-provider might be able to develop a policy that improves its own local expected utility enough to more than compensate for the loss in the requesting agent's expected utility.

Space precludes giving many of the details of our extensions to factor in the providers' local utilities, so we summarize it as follows. We introduce a new linear program that allows us to compute a lower bound in the probability space, representing the maximum probability for the commitment at a given time that still allows the provider to maximize its own local utility:

$$\max \rho_{s_k} \begin{vmatrix} \forall j, \sum_a x_{ja} - \gamma \sum_{a,i} x_{ia} P\left(j|i,a\right) = \alpha_j \\ \forall i \forall a, x_{ia} \geq 0 \\ \forall s \sum_{\{i|\{time(i)=t_s \wedge Status_s(i)=F\}\}} \sum_a x_{ia} \geq \rho_s \\ \sum_i \sum_a x_{ia} R\left(i,a\right) \geq EU^* \end{vmatrix} \qquad (5)$$

Note that this is only a slight modification of Equation 3: a constraint has been added to ensure that the expected utility of the policy is at least EU^*, the best local utility achieved by the service provider. This allows us to define a lower boundary in the commitment space (shown in Figure 6), below which commitments are guaranteed to be suboptimal.

As a result, the space of probability-time commitments worth considering is in the area between these boundaries. We can exploit discretization in the time dimension, along with finding/creating discretizations in the probability dimension (since pure policies will not allow all possible probabilities to be achievable), to further prune the candidate space. Finally, by augmenting the protocol to also exchange information about the expected utility gains/losses for (counter)proposed commitments, agents can search for a commitment that increases their collective expected reward. Our longer draft of this paper describes these issues more formally.

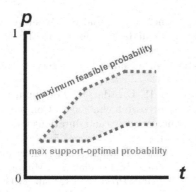

Fig. 6. A Richer Commitment Space

7   Discussion

Preliminary empirical results show that we can solve a scaled version of our example problems, finding reasonable joint policies efficiently. But a true evaluation of our approach is pending our development of a random-problem generator that includes interesting internal task structures for providing services and using service results. Our plan is to use the generator to compare our approach to a range of other approaches, ranging from naive (brute-force) multi-agent MDP techniques [13], to alternating maximization techniques like JESP [14], to techniques geared toward solving restricted aspects of the problem very well, including disjunctive-temporal-problem (DTP) planning systems [6,7].

Analytically, we have in the past shown that the space of joint policies to consider in the worst-case is substantially larger than the space of probabilistic commitments that agents can form with each other [11]. While the temporal extension to commitments increases the size of the commitment space, our initial investigations suggest that the commitment space is still much smaller, in the worst case. We expect that a more complete analysis, coupled with empirical results to tell us what happens in practice (rather than in the worst-case), will enable us to make more definitive statements about the usefulness of a commitment-based approach to service coordination.

Acknowledgments

We appreciate the comments of the anonymous reviewers. This material is based upon work supported by the Air Force Office of Scientific Research under Contract No. FA9550-07-1-0262. Any opinions, findings and conclusions or recommendations expressed in this material are those of the authors and do not necessarily reflect the views of the United States Air Force.

References

1. ter Beek, M.H., Bucchiarone, A., Gnesia, S.: A survey on service composition approaches: From industrial standards to formal methods. Technical Report 2006-TR-15 (2006)
2. Rao, J., Su, X.: A survey of automated web service composition methods. In: SWSWPC, pp. 43–54 (2004)
3. Papazoglou, M.P., Traverso, P., Dustdar, S., Leymann, F.: Service-oriented computing: State of the art and research challenges. Computer 40(11), 38–45 (2007)
4. Vere, S.A.: Planning in time: Windows and durations for activities and goals. IEEE Trans. Pattern Anal. and Machine Intelligence 5, 246–267 (1983)
5. Boutilier, C., Dean, T., Hanks, S.: Planning under uncertainty: Structural assumptions and computational leverage. In: Ghallab, M., Milani, A. (eds.) New Directions in AI Planning, pp. 157–172. IOS Press, Amsterdam (1996)
6. Tsamardinos, I., Pollack, M.: Efficient solution techniques for disjunctive temporal problems (2002)
7. Stergiou, K., Koubarakis, M.: Backtracking algorithms for disjunctions of temporal constraints. Artificial Intelligence 120(1), 81–117 (2000)
8. Kallenberg, L.: Linear Programming and Finite Markovian Control Problems. Math. Centrum, Amsterdam (1983)
9. Puterman, M.L.: Markov Decision Processes: Discrete Stochastic Dynamic Programming. John Wiley & Sons, Chichester (1994)
10. D'Epenoux: A probabilistic production and inventory problem. Management Science 10, 98–108 (1963)
11. Witwicki, S., Durfee, E.: Commitment-driven distributed joint policy search. In: Proceedings of the Sixth Intl. Joint Conf. on Autonomous Agents and Multiagent Systems (AAMAS 2007), Honolulu, HI, pp. 480–487 (2007)
12. Musliner, D.J., Durfee, E.H., Wu, J., Dolgov, D.A., Goldman, R.P., Boddy, M.S.: Coordinated plan management using multiagent MDPs. In: Working Notes of the AAAI Spring Symp. on Distributed Plan and Schedule Management (March 2006)
13. Goldman, C.V., Zilberstein, S.: Decentralized control of cooperative systems: Categorization and complexity analysis. Journal of Artificial Intelligence Research 22, 143–174 (2004)
14. Nair, R., Varakantham, P., Tambe, M., Yokoo, M.: Networked distributed POMDPs: A synthesis of distributed constraint optimization and POMDPs. In: AAAI 2005, pp. 133–139 (2005)

Author Index

Lecture Notes in Computer Science

Sublibrary 3: Information Systems and Application, incl. Internet/Web and HCI

For information about Vols. 1– 4577
please contact your bookseller or Springer

Vol. 4805: R. Meersman, Z. Tari, P. Herrero (Eds.), On the Move to Meaningful Internet Systems 2007: OTM 2007 Workshops, Part I. XXXIV, 757 pages. 2007.

Vol. 4804: R. Meersman, Z. Tari (Eds.), On the Move to Meaningful Internet Systems 2007: CoopIS, DOA, ODBASE, GADA, and IS, Part II. XXIX, 683 pages. 2007.

Vol. 4803: R. Meersman, Z. Tari (Eds.), On the Move to Meaningful Internet Systems 2007: CoopIS, DOA, ODBASE, GADA, and IS, Part I. XXIX, 1173 pages. 2007.

Vol. 4802: J.-L. Hainaut, E.A. Rundensteiner, M. Kirchberg, M. Bertolotto, M. Brochhausen, Y.-P.P. Chen, S.S.-S. Cherfi, M. Doerr, H. Han, S. Hartmann, J. Parsons, G. Poels, C. Rolland, J. Trujillo, E. Yu, E. Zimányie (Eds.), Advances in Conceptual Modeling – Foundations and Applications. XIX, 420 pages. 2007.

Vol. 4801: C. Parent, K.-D. Schewe, V.C. Storey, B. Thalheim (Eds.), Conceptual Modeling - ER 2007. XVI, 616 pages. 2007.

Vol. 4797: M. Arenas, M.I. Schwartzbach (Eds.), Database Programming Languages. VIII, 261 pages. 2007.

Vol. 4796: M. Lew, N. Sebe, T.S. Huang, E.M. Bakker (Eds.), Human–Computer Interaction. X, 157 pages. 2007.

Vol. 4794: B. Schiele, A.K. Dey, H. Gellersen, B. de Ruyter, M. Tscheligi, R. Wichert, E. Aarts, A. Buchmann (Eds.), Ambient Intelligence. XV, 375 pages. 2007.

Vol. 4777: S. Bhalla (Ed.), Databases in Networked Information Systems. X, 329 pages. 2007.

Vol. 4761: R. Obermaisser, Y. Nah, P. Puschner, F.J. Rammig (Eds.), Software Technologies for Embedded and Ubiquitous Systems. XIV, 563 pages. 2007.

Vol. 4747: S. Džeroski, J. Struyf (Eds.), Knowledge Discovery in Inductive Databases. X, 301 pages. 2007.

Vol. 4744: Y. de Kort, W. IJsselsteijn, C. Midden, B. Eggen, B.J. Fogg (Eds.), Persuasive Technology. XIV, 316 pages. 2007.

Vol. 4740: L. Ma, M. Rauterberg, R. Nakatsu (Eds.), Entertainment Computing – ICEC 2007. XXX, 480 pages. 2007.

Vol. 4730: C. Peters, P. Clough, F.C. Gey, J. Karlgren, B. Magnini, D.W. Oard, M. de Rijke, M. Stempfhuber (Eds.), Evaluation of Multilingual and Multi-modal Information Retrieval. XXIV, 998 pages. 2007.

Vol. 4723: M. R. Berthold, J. Shawe-Taylor, N. Lavrač (Eds.), Advances in Intelligent Data Analysis VII. XIV, 380 pages. 2007.

Vol. 4721: W. Jonker, M. Petković (Eds.), Secure Data Management. X, 213 pages. 2007.

Vol. 4718: J. Hightower, B. Schiele, T. Strang (Eds.), Location- and Context-Awareness. X, 297 pages. 2007.

Vol. 4717: J. Krumm, G.D. Abowd, A. Seneviratne, T. Strang (Eds.), UbiComp 2007: Ubiquitous Computing. XIX, 520 pages. 2007.

Vol. 4715: J.M. Haake, S.F. Ochoa, A. Cechich (Eds.), Groupware: Design, Implementation, and Use. XIII, 355 pages. 2007.

Vol. 4714: G. Alonso, P. Dadam, M. Rosemann (Eds.), Business Process Management. XIII, 418 pages. 2007.

Vol. 4704: D. Barbosa, A. Bonifati, Z. Bellahsène, E. Hunt, R. Unland (Eds.), Database and XML Technologies. X, 141 pages. 2007.

Vol. 4690: Y. Ioannidis, B. Novikov, B. Rachev (Eds.), Advances in Databases and Information Systems. XIII, 377 pages. 2007.

Vol. 4675: L. Kovács, N. Fuhr, C. Meghini (Eds.), Research and Advanced Technology for Digital Libraries. XVII, 585 pages. 2007.

Vol. 4674: Y. Luo (Ed.), Cooperative Design, Visualization, and Engineering. XIII, 431 pages. 2007.

Vol. 4663: C. Baranauskas, P. Palanque, J. Abascal, S.D.J. Barbosa (Eds.), Human-Computer Interaction – INTERACT 2007, Part II. XXXIII, 735 pages. 2007.

Vol. 4662: C. Baranauskas, P. Palanque, J. Abascal, S.D.J. Barbosa (Eds.), Human-Computer Interaction – INTERACT 2007, Part I. XXXIII, 637 pages. 2007.

Vol. 4658: T. Enokido, L. Barolli, M. Takizawa (Eds.), Network-Based Information Systems. XIII, 544 pages. 2007.

Vol. 4656: M.A. Wimmer, J. Scholl, Å. Grönlund (Eds.), Electronic Government. XIV, 450 pages. 2007.

Vol. 4655: G. Psaila, R. Wagner (Eds.), E-Commerce and Web Technologies. VII, 229 pages. 2007.

Vol. 4654: I.-Y. Song, J. Eder, T.M. Nguyen (Eds.), Data Warehousing and Knowledge Discovery. XVI, 482 pages. 2007.

Vol. 4653: R. Wagner, N. Revell, G. Pernul (Eds.), Database and Expert Systems Applications. XXII, 907 pages. 2007.

Vol. 4636: G. Antoniou, U. Aßmann, C. Baroglio, S. Decker, N. Henze, P.-L. Patranjan, R. Tolksdorf (Eds.), Reasoning Web. IX, 345 pages. 2007.

Vol. 4611: J. Indulska, J. Ma, L.T. Yang, T. Ungerer, J. Cao (Eds.), Ubiquitous Intelligence and Computing. XXIII, 1257 pages. 2007.

Vol. 4607: L. Baresi, P. Fraternali, G.-J. Houben (Eds.), Web Engineering. XVI, 576 pages. 2007.

Vol. 4606: A. Pras, M. van Sinderen (Eds.), Dependable and Adaptable Networks and Services. XIV, 149 pages. 2007.

Vol. 4605: D. Papadias, D. Zhang, G. Kollios (Eds.), Advances in Spatial and Temporal Databases. X, 479 pages. 2007.

Vol. 4602: S. Barker, G.-J. Ahn (Eds.), Data and Applications Security XXI. X, 291 pages. 2007.

Vol. 4601: S. Spaccapietra, P. Atzeni, F. Fages, M.-S. Hacid, M. Kifer, J. Mylopoulos, B. Pernici, P. Shvaiko, J. Trujillo, I. Zaihrayeu (Eds.), Journal on Data Semantics IX. XV, 197 pages. 2007.

Vol. 4592: Z. Kedad, N. Lammari, E. Métais, F. Meziane, Y. Rezgui (Eds.), Natural Language Processing and Information Systems. XIV, 442 pages. 2007.

Vol. 4587: R. Cooper, J. Kennedy (Eds.), Data Management. XIII, 259 pages. 2007.